THE INFINITE
JOURNEY

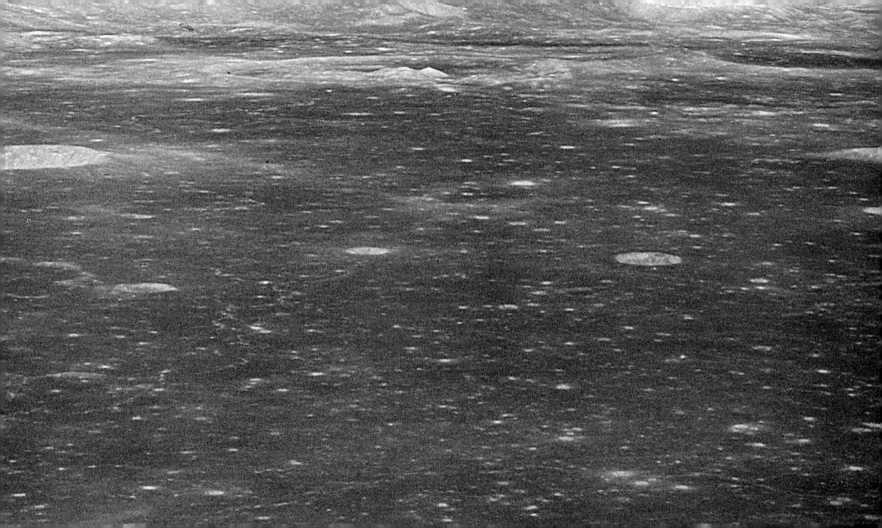

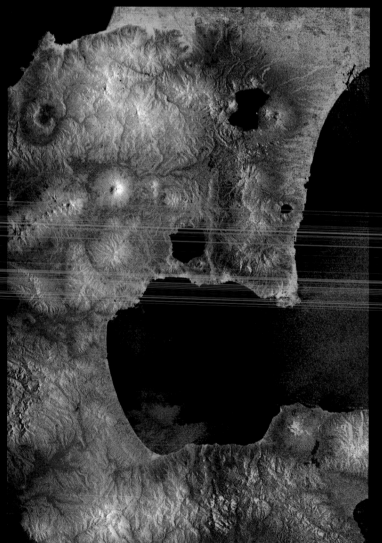

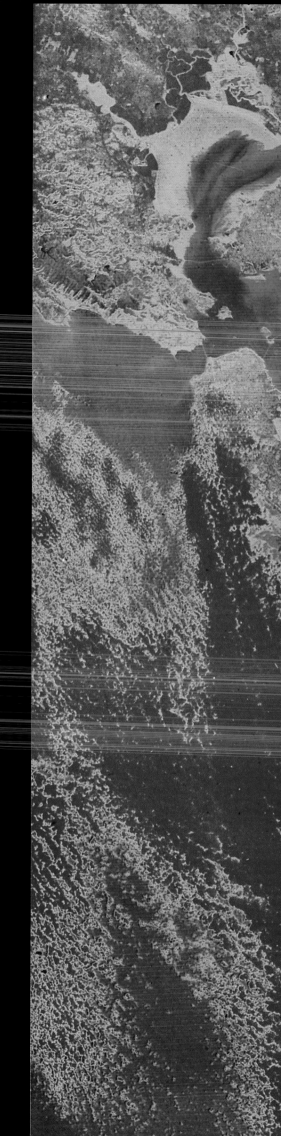

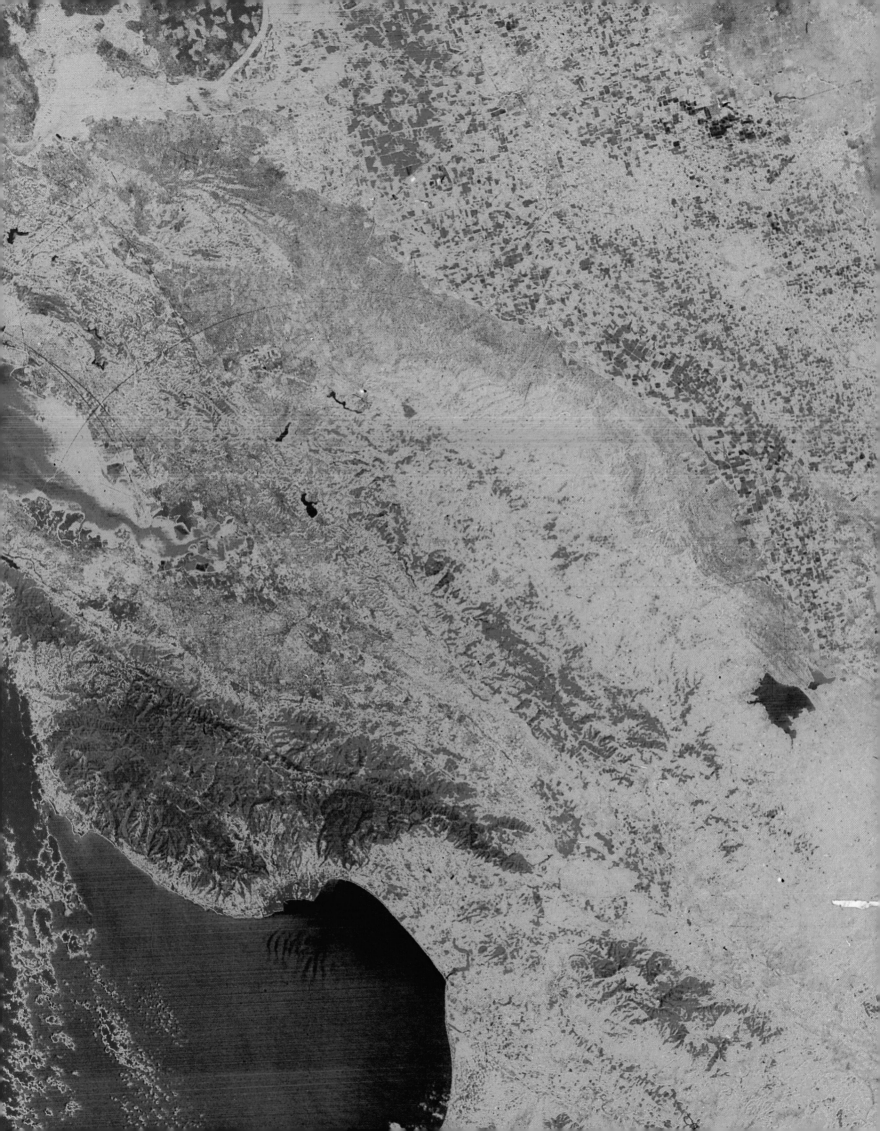

THE INFINITE

JOURNEY

Eyewitness Accounts of NASA

and the Age of Space

Page 1
This image composite was taken by the Galileo spacecraft in 1992 during its mission to Jupiter. The view of Earth includes South America and swirling storm clouds over the South Pacific. Showing brightly at the bottom of the Moon is the large Tycho impact basin.

Page 2, top
Apollo 11 command module pilot Michael Collins confers with technician Joe Schmitt during suiting operations prior to the astronaut's departure for the Moon.

Page 2, bottom
An oblique view of the lunar farside as photographed from the Apollo 11 spacecraft. The large crater in the center is International Astronomical Union (IAU) crater number 308, measuring fifty miles in diameter.

Page 3
Earthrise during the Apollo 11 mission. The lunar terrain pictured is in the area of Smith's Sea on the nearside of the Moon.

Pages 4–5
Beginning the sixty-eighth flight of the space shuttle program, *Endeavour* lifts off on March 2, 1995.

Page 6, top
An eruption of the Kliuchevskoi volcano in Kamchatka, Russia. This image of one of the world's most active volcanoes was acquired by the Spaceborne Imaging Radar-C/X-band Synthetic Aperture Radar (SIR-C/X-SAR) instrument during a 1994 space shuttle mission.

Page 6, bottom
This image of the southeast part of the volcanic island of Hokkaido, Japan, was acquired by the Shuttle Radar Topography Mission (SRTM) during a February 2000 *Endeavour* shuttle flight. Elevation is represented by color, from blue at the lowest elevations to white at the highest.

Pages 6–7
An Earth Resources Technology Satellite (ERTS) infrared image of the San Francisco Bay and Monterey Bay areas in California.

Pages 8–9
A dramatic view of Jupiter's Great Red Spot and its surroundings taken by Voyager 1 in 1979. Cloud details as small as 100 miles across can be seen, and an extraordinarily complex and variable cloud pattern appears to the left of the Red Spot.

Right
Astronaut Jerry Ross performs a space walk during the 1998 STS-88 joint U.S.–Russia mission to join Unity and Zarya, the first modules of the International Space Station. The photographer, mission specialist James Newman, is seen reflected in Ross's helmet visor.

CONTENTS

THE WHITE HOUSE

WASHINGTON

July 11, 2000

On October 1, 1958, the National Aeronautics and Space Administration came into being; designed, as its authorizing legislation stated, "for research into the problems of flight within and outside the Earth's atmosphere, and for other purposes." Those "other purposes" would beckon Americans to set off on the infinite journey that is chronicled within the pages of this extraordinary book.

I was still a student at Oxford when Neil Armstrong and Buzz Aldrin first walked on the moon. Thirty years later, on the anniversary of that historic landing, I was privileged to meet with both of them and with Michael Collins in the Oval Office. They left me a gift on loan from NASA: a golfball-sized moon rock, which I proudly show off to visitors. I've met with many others in the Oval Office since I've had that rock on the table, and sometimes the discussions—about gun control, health care, and so forth—can become heated. Voices are raised, people become agitated or angry at one another. That's when I'll tell everyone to pause and take a moment to look at that moon rock, which is 3.6 billion years old. It's a wonderful way to restore our perspective.

That rock is also a daily reminder to me of America's capacity to meet new challenges and reach unimagined heights. The men and women of NASA embody that capacity, and I am confident that, in this new millennium, their energy, creativity, and sense of adventure will continue to broaden our horizons, improve our lives, power humankind's pioneering explorations of space, and renew our sense of wonder at the universe.

Bill Clinton

Walter Cronkite

FOREWORD

We are extraordinarily lucky, we of the generation that participated in and witnessed the birth of the space age.

We were there at the concept as the military tested the first American rockets in the New Mexico desert. We were there as President Kennedy put a new focus on the space program. He inspired a nation and spoke the American dream—within the decade we would land men on the Moon and bring them safely home.

And we were there on the beaches of Florida and, the lucky few, at Cape Canaveral as we watched our rockets and spacecraft grow and eventually soar toward the Moon. And the world, through television, shared the American adventure, and the American success.

The Moon landing was dramatic and important but it was, after all, only a step along the way. The shuttle was to follow—our first reusable spacecraft, one that soon would be on the frequent schedule of a commuter as it orbited Earth, photographing Earth as it never had been shown before, revealing the stars as we had never seen them, enabling unprecedented predictions of our weather below, providing the satellite communication links that forever changed the world, aiding the nations in assuring peace by keeping a friendly eye on each other's neighbors, and proving humans' capabilities to work in the rare environment of space.

With the closing of the remarkable century that saw space fall to the conquering miracles of human flight, sixteen nations were participating in building the first space station that will be continually occupied by humans—a station that will open new possibilities for further exploration and uses of the near gravity-free environment for medical treatment and manufacturing techniques that still are only dreams.

And parallel to the accomplishments of human flight, we have lived with the exciting development of robotic exploration of deep space. Those amazing craft conceived by the best brains of humans and directed by them are today circling the distant planets at the edge of our universe and telling us what they see and feel and sense.

Our successes in space were, for Americans, a powerful antidote at a time when we needed it badly. The 1960s, when we first launched humans into space and went on to the Moon, were in other ways a terrible drain on our spirit. The civil rights battles, the frightening divisiveness of the Vietnam War, the horrible assassinations—they drained the American spirit. It is no exaggeration to say the space program saved us.

Yes, indeed, we are the lucky generation. Not only were our achievements in space important in restoring our self-respect, they enabled us as well to enter the history books. Thanks to our hero scientists, technicians, and astronauts, we will be remembered hundreds of years from now as those who first broke our earthly bonds and ventured into space. From our descendants' perches on other planets or distant space cities, they will look back at our achievement with wonder at our courage and audacity and with appreciation at our accomplishments, which assured the future in which they live.

CBS news anchor

Walter Cronkite tries out a low-gravity simulator during a 1968 visit to NASA's Langley Research Center in Virginia. Apollo astronauts used the rig to practice walking in the Moon's gravity—one-sixth that of Earth's.

Pages 14–15

The Russian space station Mir floats over the Pacific Ocean during 1995 rendezvous operations with the space shuttle *Discovery*.

William E. Burrows

INTRODUCTION

Humanity's fascination with the planets, stars, and ever-changing Moon, sailing silently across the night sky in a majestic and predictable procession, is as old as civilization. We have always wanted to know what was out there. Only in the last century of the millennium, however, did we finally bridge the gap between Earth and the firmament in which it exists. While some were scarring the century with wars and genocide, many thousands of others were ennobling humankind by sending astronauts into space and even to the Moon; studying neighboring worlds and discovering planets in distant galaxies; and looking back to the beginning of time and to the edges of the universe. To people who care about the future of the species, those thousands of scientists, engineers, technicians, astronauts, and managers are the heroes who created the twentieth century's most notable accomplishment: finally escaping Earth's bonds and beginning the endless migration that will carry future generations of men and women across the solar system and beyond.

That giant leap will follow a path formed by centuries of smaller steps. It will be a leap foretold by Daedalus, who made wings so he could escape from King Minos's labyrinth on Crete. In doing so, he combined flight with freedom and bound them with courage.

A section of the Saturn V Moon rocket leaves a Mississippi test stand in 1967 on its way to the launch-pad in Florida.

Wise men of every ancient culture studied the vast canopy of the heavens and formed deep bonds with it. Astronomer-priests, in Babylon forty-five hundred years ago, in China, in the Americas, and in Egypt, recorded the stately movement of the Moon that defined weeks, months, and years, telling them when to plant and harvest and where to seasonally fish. They meticulously studied the yearly circling of the constellations that they believed determined human fate. They built observatories, invented calendars, charted the paths of planets, witnessed eclipses with fear and awe, and gazed in wonder at comets and meteors streaking across the night sky.

The ancients were scrupulous observers of the firmament because they believed that the heavens were the home of the gods and that celestial activities shaped life on Earth: Not until the sixth century B.C.E., when travelers brought the records of Babylon and Egypt back to Greece, did astronomy move out of the realm of myth and into that of science. In the fourth century B.C.E., Aristotle described a universe centered on Earth, which he believed was surrounded by a sphere containing all the stars, moving in perfect circles and therefore in a state of perfect harmony. A century later, the less well known Aristarchus of Samos made what was undoubtedly the first model of a Sun-centered universe. It was largely ignored.

The more appealing theory of an Earth-centered cosmos would prevail for fifteen hundred years. It was most brilliantly defined in 140 C.E. by the Greek astronomer Ptolemy of Alexandria, who produced an encyclopedia of all known celestial observations and himself identified 1,022 stars and gave them latitudes, longitudes, and magnitudes. Ptolemy believed the universe was a system of nested spheres on which the Sun, Moon, and stars revolved in circular orbits at a uniform speed, making it possible to predict the motions of heavenly bodies.

Ptolemy's system held sway until the European Renaissance produced a galaxy of luminaries whose investigations disproved it. Their truly scientific observations and theories created modern astronomy and therefore led the way to space exploration. Among them, Nicolaus Copernicus was the first to reassert Aristarchus's theory; Copernicus's remarkable *On the Revolution of the Celestial Spheres* was published in 1543. He was followed closely by Johannes Kepler who, basing his calculations on the theory of a heliocentric universe, defined three laws of planetary motion that allowed him to predict the positions of the planets much more accurately than Ptolemy. Then in 1610 Galileo Galilei became the first astronomer to turn the newly invented telescope toward the heavens. Among the discoveries he made with his primitive "optick stick" was that of the four moons circling Jupiter in simple orbits—a kind of miniature of the Copernican solar system.

But Galileo saw more than Jovian moons. When he turned his optick stick toward his own planet's solitary companion, he saw the same kinds of mountains and valleys that are on Earth. If Earth was habitable, then so was the Moon. And if that was true, then the challenge was not being able to live on the Moon, but just getting there.

Kepler could not contain his excitement about the Italian's discoveries. "Who would have believed that a huge ocean could be crossed more peacefully

The Apollo 11 astronauts snapped this picture of the Moon on their way home, a day after Neil A. Armstong and Edwin E. "Buzz" Aldrin's historic first steps on the lunar surface, July 20, 1969. The two-and-a-half-hour Moon walk capped nearly a decade of work by more than 400,000 people.

and safely than the narrow expanse of the Adriatic, the Baltic Sea, or the English Channel?" he asked Galileo in a hauntingly beautiful and prophetic letter. "Provide ship or sails adapted to the heavenly breezes, and there will be some who will not fear even that void [of space]. . . . So, for those who will come shortly to attempt this journey, let us establish the astronomy: Galileo, you of Jupiter, I of the Moon." They would come, all right, but not shortly.

An eclectic fraternity had begun to pave the road to space. Astronomers and other scientists relentlessly pushed back myth and superstition and replaced them with knowledge: Less than a century after Galileo, Sir Isaac Newton published his equation-laden masterpiece *Philosophiae naturalis principia mathematica (Mathematical Principles of Natural Philosophy)*, which invented the theories of motion underlying rocketry and therefore space travel. Meanwhile, storytellers looked at the sky and saw awe-inspiring mystery and magic. The storytellers—later to be called science-fiction writers—eagerly used the science that was replacing magic. The relationship between storytellers and scientists was wonderfully symbiotic: Novelists used science to make space travel plausible and in doing so, they set long-range goals for the very scientists who inspired them.

Even Kepler wrote a fantastical novel about a trip to the Moon. It included a meeting with serpentine moon creatures but also scientific speculation about escaping Earth's gravity and surviving in the vacuum of space. Similar fantasies appeared from time to time until 1865, when Jules Verne published his epic space adventure, *From the Earth to the Moon*, a tale of three intrepid adventurers whose spaceship was shot out of the muzzle of a colossal cannon named *Columbiad*. Verne soaked up enough real science to site *Columbiad* in Florida, where there was more Earth spin than at higher latitudes, thereby helping to fling the huge cannonball into space. With this novel, modern science fiction was born.

Verne and other science-fiction writers fired the imagination of generations of scientists, who bent to the task of solving the many problems that prevented earthlings from heading to space. Chief among them was the means of getting there. A device was required that would both overcome gravity and function in an airless environment. It was called a rocket.

By the time the twentieth century dawned, the device had caught the attention of scientists who had spaceflight in mind.

The Rocket Pioneers

A rocket is an internal combustion engine that operates without outside air. It contains a chamber in which fuel and an oxidizer are burned together, producing hot gases that turn the chamber into a blast furnace, where the intensely hot, expanding gases are pushed in all directions. Some escape through an open nozzle at the bottom of the engine. The rest push against the top and sides of the combustion chamber, sending it (and the rocket in which it is bolted) in the opposite direction from the nozzle. Whether the fuel is solid or liquid the principle is the same. It is Newton's third law of motion: For every action (in this case, escaping gases), there is an equal and opposite reaction (thrust).

There were rockets long before Newton. The Chinese used them in warfare at least since the twelfth century, and they appeared in Europe soon afterward. As conventional artillery improved, military rockets were abandoned but not forgotten. Early in the twentieth century, three brilliant pioneers began to imagine their use for space travel.

▼

Verne and other science-fiction writers fired the imagination of generations of scientists, who bent to the task of solving the many problems that prevented earthlings from heading to space. Chief among them was the means of getting there.

In orbit at last: Rocket builder Wernher von Braun (right), scientist James Van Allen (center), and Jet Propulsion Laboratory director William Pickering celebrate the launch of America's first satellite, Explorer 1, on January 31, 1958.

The early days of spaceflight saw more misses than hits. Pioneer 1, the first and one of the few Air Force lunar missions, tried but failed to reach the Moon in October 1958.

The first was Konstantin E. Tsiolkovsky, a self-educated Russian school-teacher and mathematician born in 1857. Tsiolkovsky was a self-described gravity hater who believed that the spirits of everyone who had ever lived were out in space looking for a home. He was also captivated by Verne's great novel. He therefore applied mathematics and physics, including Newton's three laws of motion, to invent a utilitarian machine that could reach the precincts of all the world's departed souls: a liquid-fueled rocket. Tsiolkovsky published his first paper on the subject in 1903 and, in the quarter century that followed, advocated various fuels (chiefly liquid oxygen and hydrogen, which are light and powerful), steering devices, and even "step" rockets, in which several rocket stages fall away after they burn out, reducing deadweight. These, he thought, could produce the high velocity that is necessary to overcome gravity and get to space.

Perhaps because he lacked resources for experiments, Tsiolkovsky's work was all theoretical. It took another devotee of Jules Verne to turn theory into reality. He was the American physicist Robert H. Goddard, who in 1919 wrote a seminal paper, later published, for the Smithsonian Institution: *A Method of Reaching Extreme Altitudes*. The sixty-nine-page document described in detail how to build a two-stage, solid-propellant rocket that could carry scientific research instruments to very high altitudes to measure weather and collect other data, after which they would parachute gently back to Earth.

The *New York Times* ridiculed Goddard's physics (incorrectly), and other newspapers contemptuously called him the "Moon Man" because he had the temerity to claim that a rocket could reach the Moon. This made the ordinarily shy professor a recluse. But it did not stop his work. On the afternoon of March 16, 1926, he assured his place in history by successfully launching the first liquid-fueled rocket. In the years that followed, working in seclusion in Roswell, New Mexico, he continued to explore rocketry, wrote the second of his great papers, *Liquid-Propellant Rocket Development*, and registered 214 patents for rocket components.

The third of rocketry's Holy Trinity was Hermann Julius Oberth, born in 1894. Oberth was a German who combined what he learned in medical school in Munich with a longing for space travel to describe spaceflight and its effect on the human body. His 1923 study, *The Rocket into Interplanetary Space*, was loaded with calculations and descriptions covering the essentials of spaceflight in considerable detail. To rocket cognoscenti in the Germany of the mid-1920s, Oberth was a hero and his ninety-two-page tract a bible. The German enthusiasts, thrilled at the rocket's potential for finally breaking the bonds of Earth and getting people to space, formed their own association. In the Soviet Union, the Group for the Study of Reaction Motion, begun in Moscow and Leningrad, soon had chapters throughout the country. The members of all these rocket clubs were amateurs, of course, but since there were no professionals, they were rocket science's cutting edge.

Of all the clubs, Germany's led the world. The *Verein für Raumschiffahrt* (Society for Space Travel), best known as the VfR, was a group of dues-paying, workaday dreamers who launched rockets in a Berlin suburb and celebrated themselves with black-tie dinners. Hermann Oberth was a member, but one of his protégés, a young German baron from Pomerania, soon outshone the rest. He was an engineering student named Wernher von Braun.

This young man's chance for fame arose as Germany secretly began to rearm following its defeat in the First World War. The Treaty of Versailles prohibited

▼

The *New York Times* ridiculed Goddard's physics (incorrectly), and other newspapers contemptuously called him the "Moon Man" because he had the temerity to claim that a rocket could reach the Moon. This made the ordinarily shy professor a recluse. But it did not stop his work.

▼

The world was changed

forever. Sputnik vaulted
space, ending forever
Aristotle's two-dimen-
sional universe, and
beckoned men to follow.

German development of traditional weapons such as warships, warplanes, armor, and artillery, but not of rockets, because those who drafted the document were ignorant about them. Officers of the German army had a vision, however: They saw rockets, with their great range, as unparalleled artillery weapons. They therefore approached the VfR as early as 1932 and proposed to subsidize the group's research, with the understanding that the rocketeers would secretly develop rockets not to hit Mars but to hit their nation's enemies.

It was an offer the rocketeers could not refuse. Their research produced the V-2 ballistic missile, first flown by Nazi Germany in 1942. The forty-six-foot-long rocket, fueled by alcohol oxidized by supercold liquid oxygen, flew at speeds greater than thirty-five hundred miles per hour and delivered a twenty-two-hundred-pound warhead. There was no defense against it, and by the end of the war it had demonstrated the destructive capability of the ballistic missile in Belgium, France, and most notably in England.

This rocket's potential as a powerful, strategic, postwar weapon was wasted on no one, least of all the nations uneasily allied against Nazi Germany. Leaders in the United States and the Soviet Union knew throughout the war that they were on a collision course. And many of their soldiers and scientists knew that missiles would play a significant role in the coming conflict.

At World War II's end, the U.S. and Soviet victors therefore ransacked Germany of every V-2 they could lay their hands on. The U.S. Army grabbed about one hundred unassembled missiles in the Soviet occupation zone, virtually under the eyes of the Russians. The Americans also collected all of the German rocket team's best scientists and engineers, led by von Braun, and quickly whisked them off to White Sands, New Mexico. There von Braun and his group tested their rockets as ballistic missiles. And, of course, they shared information about the basic design with the embryonic aerospace establishment.

The scraps that were left behind in Germany included a handful of V-2s, a few thousand technicians, and some of the engineering group's lesser lights. These were appropriated by the Russians and transported to the Soviet heartland. There, Tsiolkovsky's heirs—fabled Chief Designer Sergei P. Korolyov, rocket-engineering genius Valentin Glushko, all-purpose prodigy Mikhail Tikhonravov, and others— used the V-2 to develop the U.S.S.R.'s own rockets. The goal was to create a missile that could carry a six-ton hydrogen warhead over the top of the world to obliterate targets in America. That rocket, called the R-7, was finally ready in 1957. But carrying H-bombs was not to be its claim to fame.

The Dawn of the Space Age

At four minutes to midnight on October 4, 1957, at the new Tyuratam rocket facility in the Soviet Socialist Republic of Kazakhstan, an R-7 roared off the launchpad carrying a 184-pound satellite named Sputnik, or "fellow traveler," into the eastern sky. Within five minutes, the first man-made objects—the rocket's spent main stage, the shroud that protected the satellite as it rose through the thinning atmosphere, and the satellite itself, now sprouting four spring-loaded antennas—were sailing over North America in celestial formation.

The world was changed forever. Sputnik vaulted over the boundary between Earth and space, ending forever Aristotle's two-dimensional universe, and beckoned men to follow. Mikhail Tikhonravov, a truly unsung hero and visionary who had conceived both the R-7's rocket cluster and the spacecraft itself, allowed himself a little hyperbole. "This night," he recalled thinking, years later, "has become one of the most glorious in the history of humanity."

It wasn't glorious for the United States. Americans had come to regard their former ally with fear and loathing. Most disapproved of Communist philosophy in general and Communist aggression in particular. The Soviet Union's absorption and brutal occupation of most of Eastern Europe were seen as a menacing scourge. So were the Communist victory in China in 1949, the testing of a Soviet atomic bomb the same year, the attack on South Korea in 1950, and the spread of insurrections from Greece to the Philippines.

The United States liked to consider itself the guardian of Western democracy and the planet's supreme technological innovator. It reckoned its Communist opponent to be a scientific sloth. But Sputnik changed all that, literally overnight. Senate majority leader Lyndon B. Johnson, a former Texas schoolteacher, was soon to become a driving force behind the fledgling space program. He captured America's doom-laden mood in a few words. Romans once controlled the world because they could build roads, he drawled. Later the British dominated the seas with ships and later still the Americans used planes to control the air. "Now the Communists have established a foothold in outer space. It is not very reassuring to be told that next year we will put a better satellite into the air. Perhaps it will also have chrome trim and automatic windshield wipers."

Johnson's final remark was a swipe at the consumer society of the booming 1950s, preoccupied with tail-finned automobiles and other frivolous toys. In painful contrast, Sputnik seemed to prove that the Russians were a tough and dedicated people whose totalitarian leaders knew how to harness institutional self-sacrifice to achieve a higher goal. Although President Dwight D. Eisenhower insisted, correctly, that the Soviet Union had no meaningful lead in science and technology, American politicians, scientists, and the press all leaped on the contrast between their soft and hedonistic compatriots, typified by the comic-book characters Archie, Jughead, Betty, Veronica, Reggie, and other self-absorbed, intellectually listless teenagers, and their disciplined adversary.

Their anxieties seemed confirmed barely a month later on November 3, when Sputnik 2 went up. This satellite was bigger than its predecessor and carried a passenger: a dog named Laika. A dog meant that the Russians were testing for effects of radiation and zero gravity on living organisms, with a view toward sending a man to orbit Earth.

And whatever the president might say, no one was reassured on December 6, when the United States attempted its first launch of a scientific satellite from Florida's Cape Canaveral. The launcher was a U.S. Navy Vanguard rocket, a sleek thoroughbred carrying a grapefruit-sized orbiter of the same name inside its shroud. In full view of hundreds of reporters and television spectators from coast to coast, the rocket erupted into a ghastly orange and black fireball. Russian premier Nikita S. Khrushchev reportedly quipped that the American entry would have been more accurately named "rearguard." Mortified newspaper editors called it "Kaputnik."

The U.S. Army and the California Institute of Technology's Jet Propulsion Laboratory in Pasadena finally retrieved the nation's prestige when on the night of January 31, 1958, a U.S. Army Juno rocket carried the laboratory's Explorer 1 into orbit. As the president had promised, the American spacecraft was superior to its primitive Russian counterpart: Explorer 1, unlike Sputnik, carried scientific instru-

A V-2 rocket lifts off at White Sands, New Mexico, 1946. The German rocket scientists who came to the United States after World War II had learned their craft building such vehicles. Their expertise gave Americans the edge in the space race with the Soviets.

▼

Despite the vehement arguments of the military services, President Eisenhower was determined to make the U.S. space effort a civilian one, even knowing, as he did, that the basic vehicle for getting to space remained the ballistic missile and that the first Americans in space would be military test pilots.

ments. The one that would become most famous was a cosmic-ray detector, the brainchild of University of Iowa physicist James A. Van Allen. It documented two doughnut-shaped radiation belts that circle Earth. The Van Allen belts, as they were quickly named, provided the first confirmation that the planet is surrounded by a magnetosphere that shields it from deadly solar radiation. The discovery provided a hint of the wonders that lay ahead.

It was clear that the space age had begun in earnest and that scientific competition with the Soviet Union was as much a reality of the Cold War as military rivalry. To win the competition, the United States needed a dedicated agency to run a variety of space programs. It was time to focus America's formidable resources on the space race. So early in 1958, in an atmosphere of crisis set off by Sputnik, the president, the Congress, and the fledgling aerospace establishment swung into action.

NASA Is Born

When congressional investigators looked at American progress in aerospace studies, they found military programs for ballistic missiles and for sending airmen to space, as well as a variety of embryonic civilian scientific programs. But there was no overall supervision. The agency closest to that role was the National Advisory Committee for Aeronautics (NACA). NACA, a small, elite group of scientists and engineers, had been around since 1915. After World War II it had begun examining space-related problems. But mostly, it was known for excellent aeronautical research (its symbol was the Wright Flyer). And NACA was civilian.

This last was important. Despite the vehement arguments of the military services, President Eisenhower was determined to make the U.S. space effort a civilian one, even knowing, as he did, that the basic vehicle for getting to space remained the ballistic missile and that the first Americans in space would be military test pilots. A civilian agency's space programs could be highly publicized, showing off the virtues of democracy to the world. And while the civilian space race would provide attractive propaganda, it wouldn't be an arms race. On the most fundamental level, however, it was a charade, since civilian and military space technology were basically the same. Every civilian program had a military counterpart, with many designers wearing both "white hats" (civilian) and "black hats" (top secret). Spacecraft that would observe Earth, for example, would be called remote-sensing satellites when they did civilian work and reconnaissance or surveillance satellites when they spied on other nations.

Within six months, Congress created the National Aeronautics and Space Administration, which absorbed the venerable NACA along with a number of other programs. President Eisenhower swore in the new agency's first administrator, T. Keith Glennan, on August 19, 1958, and the establishment known ever since as NASA was formally born. Glennan proceeded to build an empire, far-flung because America's budding space assets were spread around the country, and also because the greater the number of states that had a stake in NASA's success, the more protection the fledgling agency would have in Congress.

For its major rocket launch site, NASA went to the state favored by Jules Verne a century before: Florida. The agency took possession of eighty thousand acres of the Air Force's Atlantic Missile Range at Cape Canaveral, almost due east of Tampa on the Atlantic, and began building a series of launch complexes that nearly touched the sparkling white beaches. This became the John F. Kennedy Space Center after the president's assassination in 1963.

The missile range was not the only military appropriation, however. Despite the generals' vehement objections, NASA also took control of the U.S. Army Ballistic Missile Agency at Huntsville, Alabama, where von Braun and the rest of the old rocket team were designing the Redstone, Jupiter, and other ballistic missiles, plus huge space rockets, for the U.S. Army, thus putting the service out of the long-range missile and space business. Glennan renamed it the George C. Marshall Space Flight Center, after the general who created the plan to reconstruct postwar Europe, an achievement that made him the only professional soldier ever to win the Nobel Peace Prize.

A number of new NASA research centers—appropriated or created—would specialize in critical areas of spaceflight. Among them was the Goddard Space Flight Center at Greenbelt, Maryland, which was made responsible for unmanned Earth orbit missions, most of them scientific. The Langley Research Center in Tidewater, Virginia, concentrated on advanced aeronautics and also on some astronautics: It would create the lunar orbiter that scouted likely landing places on the Moon. Wallops Island, also on the Virginia coast, worked on relatively inexpensive sounding rockets for atmospheric and other research. The Lewis Research Center in Cleveland, Ohio, managed chemical and electronic engine development, investigated nuclear and solar power, and developed scores of related devices, including propellants, pumps, and turbines. In California, the Ames Research Center near Sunnyvale covered a wide area, ranging from advanced aeronautics to studying solar physics and planetary environments to guidance, control, and life-support systems for spacecraft. In Pasadena, Caltech's Jet Propulsion Laboratory would abandon army missile work to become the world leader in planetary exploration, sending robotic explorers to the far reaches of the solar system and beyond.

There were facilities specializing in electronics research, flight-testing manned aircraft, and assembling huge rockets; and of course, there were centers devoted to the coming manned missions. Chief among them was the Space Task Group, soon renamed the Manned Spacecraft Center, in Houston, Texas, later renamed the Lyndon B. Johnson Space Center in 1973 in recognition of Johnson's championship of the space program (and his state's role in it). Houston would train astronauts and manage all manned missions from Mercury to the International Space Station.

In addition, NASA depended on universities, contractors, and subcontractors. As a result, vast numbers of people contributed to the space program, directly and otherwise. It has been estimated, for instance, that the first manned program, Mercury, employed two million. One in fifty Americans was said to have been involved in some way with Apollo. NASA was more than a group of scientific and engineering programs. It was to become its own culture.

Manned or Unmanned?

As NASA's widespread infrastructure came together, so did its programs, which fell under two fundamental, often competitive, headings: manned flight that sent people to space and unmanned flight devoted strictly to science and so-called applications such as meteorology and communication. Even before the space agency was born, the factions went to war over fundamental philosophical and professional differences (not to mention slices of a limited budgetary pie).

The manned program had originated before the formation of NASA, as a

Past meets future: A mule-drawn plow digs trenches for piping near a Saturn rocket test stand in the 1960s. The Moon landings took place before the handheld calculator was invented, and long before desktop computers.

The original seven

Mercury astronauts
(from left: M. Scott
Carpenter, L. Gordon
Cooper Jr., John H.
Glenn Jr., Virgil I. "Gus"
Grissom, Wally M.
Schirra Jr., Alan B.
Shepard Jr., Donald K.
"Deke" Slayton) were
drawn from an elite
group of test pilots
accustomed to risk and
the complexity of new
flying machines.

reaction to the shock of Sputnik. Various military services and research agencies were inventing ways to get an American into orbit ahead of the Soviets, the most notable being the Air Force Man in Space Soonest project, which had the unfortunate acronym MISS. With the formation of the space agency, the Space Task Group, composed of civilian engineers and planners and military representatives, came together to start a serious manned space program. The idea was to launch one man at a time in capsules that would orbit Earth and come back down under parachutes. The project was briefly called "Astronaut," until Abe Silverstein, NASA's well-read director of spaceflight development, came up with the more evocative name, Mercury, after the messenger god of Roman myth.

Advocates of this program believed that a human presence in space was a manifestation of the urge to explore, and that people belonged wherever they could go. More pragmatically, they liked to make the point that the space program depended on funding from Congress, Congressmen listened to their constituents, and the vast majority of their constituents didn't understand science and even feared it. But people on terra firma, they maintained, could readily identify with people—heroes—flying high over it. "They don't give ticker-tape parades for robots," as NASA official Franklin D. Martin put it.

Opposition to this view came from advocates of the unmanned program, which had originated in the imaginations of James Van Allen, of radiation-belt fame; the brilliant Hungarian-born aerodynamicist Theodore von Kármán; world-renowned physicist Lloyd Berkner; and other scientists and engineers. At bottom, they believed implicitly that rockets were a scientific tool, that the business of science was learn about the world, and that sending people to cavort in space—acrobats on a flying trapeze—was a wasteful distraction. They were impatient to get above the atmosphere to accomplish a broad range of missions, from understanding the environment around Earth to examining the other planets and their majestic retinues of moons, all of them other worlds in their own right.

Standing with them were von Braun, Korolyov, and their colleagues. Having earned their keep by providing ballistic missiles for the politicians and generals, they now felt free to push for the exploration of the solar system itself, and were joined by thousands of astronomers, geologists, and physicists.

But the ticker-tape argument prevailed: The lion's share of NASA's annual budget—80 to 85 percent—went to the manned program. (The budget balance was roughly the same in the Soviet Union, which deeply angered scientists there.) The space agency's core mission evolved into one of creating very big engineering projects that sent people to space: Mercury, Gemini, Apollo, the space shuttle, the space station, and planned expeditions back to the Moon and on to Mars.

Reaching for the Moon

Even as NASA strained to get Americans into orbit, the Russians scored again. With a shout of *Poyekhali*—"Here we go!"—a young Soviet fighter pilot named Yuri A. Gagarin thundered across the Kazakh steppe and onto the pages of history. On April 12, 1961, he became the first person to reach space.

His nation's premier, Nikita Khrushchev, was no space fanatic. But he was an exceedingly wily politician who, no less than his American counterparts, realized that proficiency in space—and certainly establishing firsts there—not only was good for the nation's morale but was relatively cheap for the propagandistic bang it delivered (unlike, for example, cancer research). He therefore kept pressing Korolyov to set still more records. The chief designer obliged by sending Gherman S. Titov to space for a record-breaking twenty-five hours and eighteen minutes in August 1961. Valentina Tereshkova became the first woman in space in June 1963. Korolyov staged the first multiperson flight when three cosmonauts were stuffed into a space capsule for a daylong mission in October 1964. And Aleksei A. Leonov became the first human to take a "space walk" when he left his *Voskhod 2* capsule at the end of a tether for ten minutes in March 1965. Left unreported was the fact that he nearly died trying to climb back into the thing.

Gagarin's flight aroused a deeply competitive American spirit. Less than a month later the balance began to shift. On May 5, 1961, Alan B. Shepard Jr. lifted off from Canaveral on top of one of von Braun's Redstone rockets and rode his *Freedom 7* Mercury capsule on a 303-mile suborbital flight. The achievement was roundly celebrated at a ceremony at the White House and elsewhere. But comparisons were dispiriting: Gagarin's spacecraft weighed almost five times as much as Shepard's. The Russian had been weightless almost seventeen times as long as the American. And beyond all that, most obviously, the fact remained that the first human to circle the world was a Communist.

But only twenty days later, with his country having logged barely fifteen

John F. Kennedy,

shown with astronauts Glenn and Shepard and NASA flight director Christopher Kraft, set the nation's sights on the Moon, but barely lived to see the last Mercury flight in 1963.

The Saturn V rocket towers above its mobile launch platform prior to the Apollo 12 launch in September 1969. The Saturn was engineering on a massive scale—it stood more than 100 yards tall and weighed 3,300 tons.

▼

As if getting to the Moon

in a scant nine years were not challenge enough, the famously competitive young president also declared that the most daring and dangerous feat in the history of human exploration would be done in full public view. His strategy was brilliant.

minutes at the fringe of space, President Kennedy challenged the nation to send Americans to the Moon and back within a decade. His dramatic speech on May 25, 1961, left no doubt that the space race was a weapon in the Cold War: "If we are to win the battle that is going on around the world between freedom and tyranny, if we are to win the battle for men's minds," he said, "the dramatic achievements in space which occurred in recent weeks should have made clear to us all, as did the Sputnik in 1957, the impact of this adventure on the minds of men everywhere who are attempting to make a determination of which road they should take. . . ."

As if getting to the Moon in a scant nine years were not challenge enough, the famously competitive young president also declared that the most daring and dangerous feat in the history of human exploration would be done in full public view. His strategy was brilliant. The announcement not only repudiated the Kremlin's secrecy but risked something the Kremlin would never dare risk: publicly showing astronauts dying in a horrifying explosion or in a crash or being stranded on another world like a pair of latter-day Robinson Crusoes. Democracy could not lose. Televising a successful liftoff and landing would decisively prove American technological (and political) superiority once and for all. On the other hand, showing a gruesome catastrophe would not only underscore the importance of a free press but in the process prove that Western democracy was resilient enough to withstand such a horrible spectacle.

The president's announcement was certainly dramatic, but it was not as impetuous as it appeared to be. No one in NASA seriously believed that the United States would go to the expense and trouble of sending astronauts up in Mercury spacecraft only to end the manned program. Wading into the surf of the "new ocean," as Kennedy poetically called outer space, would have to end in a daring, long-distance swim to the island that circled Earth. Otherwise, getting wet in the first place would make no sense.

In fact, NASA had scripted the whole operation before the famous announcement. Mercury was to be succeeded by two-man orbital missions named Gemini, for Castor and Pollux, the twin stars in the third constellation of the zodiac. Gemini would pave the way for Apollo. Furthermore, after JFK. beseeched Lyndon Johnson to come up with some way to trump the Russians, the vice president put Robert S. McNamara, the secretary of defense, and James E. Webb, the new NASA administrator, among others, on the problem. There had been unanimity, not only on the technological feasibility of sending humans to the Moon, but on its political importance as well. As McNamara and Webb put it, nonmilitary lunar exploration was "part of the battle along the fluid front of the Cold War."

Now the battle progressed in its preliminary phases, Mercury and Gemini. John H. Glenn, a member of the newly invested silver-suited knighthood that was charged with carrying the nation's colors to the Moon, became the first American to orbit Earth on February 20, 1962, when he did so in *Friendship 7*. The Mercury program ended with L. Gordon Cooper's twenty-two-orbit marathon in 1963.

The Gemini program got off the ground in 1965, when Virgil I. "Gus" Grissom and John W. Young flew one of its craft for three orbits. Two of their successors extended the mission to the four days it would take a spacecraft to reach the Moon and did the first American space walk. Gemini 12, the last in the series, carried James A. Lovell and Edwin E. "Buzz" Aldrin to fifty-nine orbits and three space walks in 1966.

By then, Apollo itself was moving forward at full throttle, with thousands

The Apollo 11 command module in lunar orbit, as seen from the lunar lander. Astronaut Michael Collins circled alone while his two crewmates made the risky descent to the surface.

of people straining to meet its staggering deadline requirement set by the assassinated president. Getting to the Moon required a complex spacecraft: It was a "stack" consisting of a command module holding the astronauts, a service module, and a two-part lunar module to get its humans onto the lunar surface and then off. A monster rocket, devised by von Braun's gang and called Saturn, would lift this assemblage into space and fling it toward the Moon.

And there was much, much more. The machines and the mission itself, which involved sending one moving object (the stack) from another moving object (Earth) to a third moving object (the Moon) with unerring accuracy, were the technological equivalent of a quarterback rolling out and pitching a perfect pass to a moving receiver several hundred miles away. It was a masterpiece of engineering and phenomenally precise navigation.

And like all space missions, Apollo was stalked by potential calamity, as happened on the launchpad on January 27, 1967, when Gus Grissom, Edward H. White, and Roger B. Chaffee were killed in a flash fire that swept their command module. Yet twenty-three months later, the program took off, first with unmanned, then with manned practice flights in Earth orbit. On Christmas Eve of 1968, Apollo 8, carrying three astronauts, circled the Moon.

Photographs of the earth taken by the Apollo 8 astronauts dramatically showed for the first time that the planet itself was a solitary spaceship, a fragile and vulnerable green, blue, and white life-support system sailing through a vast black void. Such pictures became the icons of the environmental movement and took their place among the most important images ever recorded.

Two more orbital flights—around Earth and the Moon—succeeded Apollo 8, as NASA perfected the technique to invade the Moon. Then on July 16, 1969, many thousands who wanted to be part of the greatest voyage since Columbus set sail for the New World gathered on the Florida coast. They were there to watch Apollo 11, carrying Neil A. Armstrong, Michael Collins, and Edwin E. "Buzz" Aldrin, blast off for the Moon. Four days later, with Collins circling, Armstrong and then Aldrin climbed out of the lunar module they had christened *Eagle* and set foot on the lunar surface. "That's one small step for [a] man," the first earthling to touch another world proclaimed, "one giant leap for mankind." President Richard Nixon called the period of their flight "the greatest week in the history of the world since Creation."

After the Giant Leap

Even before the Apollo program achieved its spectacular goal, funding for NASA had begun to decline, and public interest after the first Moon landing declined as well. The mood of the nation changed during the 1960s. America was embroiled in an unpopular and frustrating war in Vietnam and a convulsive attempt to correct its own civil and social problems at home. It was a decade marked by radical student unrest, an often vicious and sometimes murderous civil rights struggle, and horrifying political assassinations. The war and the new social programs absorbed federal monies, some of them coming from NASA. By 1970, President Nixon had instructed the agency to reduce its yearly budget from $4 billion to $3 billion. Apollo had been a success, he maintained. What more did the United States have to prove?

Lack of funds meant that although twenty Apollo missions had been planned, only six followed Apollo 11, with the last one being Apollo 17 in December 1972. One of them, Apollo 13, nearly ended in tragedy when a blown oxygen tank almost killed its three astronauts. Twelve men in all walked on the Moon, leaving behind devices to detect lunar quakes and other phenomena, and bringing back more than eight hundred pounds of rock and soil samples.

To make use of remaining Apollo funds and equipment and keep its hand in manned missions, NASA launched two new and short-lived programs. These were the first steps toward realizing an old dream that had been spelled out in detail by Wernher von Braun in a series of articles in *Collier's* magazine from 1952 to 1954. Von Braun envisaged an orbiting space station, built and maintained by astronauts ferried on reusable shuttles. Such a station would permit a long, ongoing program aimed at successive and interrelated goals in space science and exploration.

The first of NASA's efforts in this direction was initially called the Apollo Applications Program and then simply Skylab. It was a small orbiting science laboratory that went up in 1973 and had three crew changes before it was permanently abandoned because the agency wanted to channel precious funds elsewhere.

The second was the first evidence of détente in space. American and Soviet military and intelligence establishments might focus on war preparation, but a number of scientists on both sides kept contact at meetings, through journals,

Apollo 17, 1972:

Lunar module pilot Harrison H. Schmitt closes out the Moon program in a moment of American triumph.

Saturn as seen in false

color by Voyager 2 in
1981. The Voyager
missions revealed the
wonders of the outer
solar system for the
first time.

and by teaching at each others' universities. Many space scientists had wanted cooperative programs from the start. That was the genesis of the Apollo-Soyuz Test Project, in which still other leftover Apollo command and service modules and their Soviet Soyuz counterparts were separately launched and then docked in orbit on July 17, 1975. The historic mission ended less than a week later, when the Apollo command module splashed down in the Pacific Ocean and Soyuz landed on the steppes of central Russia.

The projects were forerunners of more ambitious efforts. Even as they were launched, NASA was working on "routine access to space"—a shuttle that could be flung into orbit to permit astronauts to work on scientific (or commercial) projects and even build a space station, and then swoop back to Earth and land like an airplane. Billed as an economical program that could pay its own way with as many as fifty-five flights a year, the shuttle actually was a project of such complexity and expense that it took twelve years to complete. *Columbia*, its first orbiter, finally roared into space on April 12, 1981, the twentieth anniversary of Gagarin's flight.

Only twenty-four shuttle flights took place between *Columbia*'s inaugural journey and *Challenger*'s hideous, fiery end on January 28, 1986. In the years that followed, NASA regrouped and moved ahead with the manned program as well as a series of bountiful science and applications missions.

Thanks to a new political atmosphere following the breakup of the Soviet Union and the end of the Cold War, the program would become a symbol of international cooperation.

By 1986, the Soviet Union had orbited its third-generation space station, named Mir, or "peace." Lessons learned on the Apollo-Soyuz project began to pay off in 1995, when the shuttle *Atlantis* docked with Mir, the first of nine such missions. And in December 1998, as the century wound to a finale, the crew of the shuttle *Endeavour* connected the first two sections of the International Space Station: Unity, built by the United States, and Zarya, or "sunrise," built and orbited by Russia.

Space Science Comes of Age

Although human spaceflight received the lion's share of attention and support during the first forty years of the space program, both applied and theoretical science flourished in the wake of James Van Allen's pioneering 1958 Explorer experiment to an extent that had been almost unimaginable before the space age. And nearly all of the science was done by robots, some of them circling Earth, others ranging across the solar system and beyond.

Within two decades of Sputnik, Earth looked like a beehive swarming with satellites; by the late 1980s there were seven thousand of them, and they have already profoundly changed life on Earth. Some of the spacecraft relay radio, telephone, and television signals. Some are part of a global-positioning system that has already begun to profoundly change the way we navigate around the planet. Others collect detailed data on Earth's landmass and oceans and study the space around the planet itself: More than fifty Explorer satellites followed Van Allen's through 1975, collecting a vast and varied amount of information about Earth's environment with advanced imaging systems and sensors. Between 1964 and 1969 they were joined by six Orbiting Geophysical Laboratories. The first Earth Resources Technology Satellite, later named Landsat, was launched in 1972 to

Fixing the Hubble

Space Telescope,
December 1993. The
space shuttle has
made working in orbit
a reality, even if the
setting remains other-
worldly.

▼

Every day during its

near encounters with the planets, and even after several of its key components became disabled, Voyager 2 whispered back a continuous stream of data that kept its science teams enthralled.

monitor and inventory land use around the world. Using multispectral cameras and flying at an altitude of more than five hundred miles, these remote-sensing spacecraft scrutinize more territory in a single day than a squadron of airplanes could in a year. There are satellites to monitor the weather and to study the interactions of the Earth's oceans, continents, islands, and atmosphere. (Their close cousins, the supersecret military reconnaissance satellites, also work night and day to keep their digitized eyes on potential threats.)

Other Earth huggers, including the Cosmic Background Explorer (or COBE), the Hubble Space Telescope, and X-ray satellites, to name only three types, fundamentally changed our view of the universe and our place in it. As the physicist Freeman Dyson observed, X-ray astronomy has shown the universe to be a dynamic and violent place, ending forever the eternally peaceful, orderly, and serene world described by Aristotle. Incredible Hubble, faithfully circling above Earth's distorting atmosphere since 1990, has changed humanity's conception of what's happening out there: Its myriad discoveries have created an experience bordering on the religious. If the images sent back to Earth by Hubble were put on computer disks and stacked, they would tower more than two miles high.

The far-ranging cousins of such Earth-orbiting satellites are even now dramatically altering our views of space and time. Only two and a half years after Sputnik, an American spacecraft named Pioneer 5 went into orbit around the Sun and radioed back data. The exploration of the rest of the solar system started in December 1962, when a spacecraft named Mariner 2 became the first emissary from Earth to visit another world. It flew past Venus in midmonth and sent back data on that mist-shrouded planet's heavy atmosphere and blistering heat. A descendant, Mariner 4, became the first spacecraft to visit Mars when it skimmed six thousand miles above the Red Planet on July 14, 1965, and sent back pictures. Successive explorers relayed ever-more detailed information as they sped past Mars and Venus.

Such "flybys," as NASA termed them, provided enough information so that later Mariners could orbit the planets and search them in much more detail. The information from orbiters such as Mariner 9, which scouted Mars, in turn was used to created detailed maps that allowed NASA to plan landings for the two Viking spacecraft that alighted there in the summer of 1976. They also paved the way for the Pathfinder mission of July 1997. And a powerful radar mapper named Magellan, launched toward Venus in 1989, began peering through the planet's carbon-dioxide veil the following year, revealing amazing lava domes, impact craters, and other features in unprecedented detail.

The most spectacular planetary mission and the greatest feat of exploration in all history was aptly called the Grand Tour. It came together when a bright young scientist calculated that in the late 1970s, Earth and the outer planets would be in a staggered line on one side of the Sun like horses on a racetrack. The formation occurs only once every 176 years, and it permitted the observation of all the planets except Pluto on a single flight.

The tour began in 1977, when NASA sent two spacecraft, Voyagers 1 and 2 toward Jupiter and Saturn. Both explorers used the powerful gravity of successive planets to pick up speed relative to the Sun, flinging themselves past Jupiter in 1979 and Saturn in 1980 and 1981. In the process, they returned reams of data on the two huge planets' atmospheres, magnetospheres, and other properties, including Saturn's complex and breathtakingly beautiful ring structure.

Probing still deeper into the solar system, Voyager 2 reached Uranus and Neptune and their attending moons in 1986 and 1989, respectively. Every day during its near encounters with the planets, and even after several of its key components became disabled, the singular spacecraft whispered back a continuous stream of data that kept its science teams enthralled.

Each encounter was described in the media around the world, and each became a kind of love-in that was part scientific revelation, part tribute to the joys of exploration, part reunion, and part celebration by people who knew they were privileged to see whole worlds no mortal had ever seen before. Voyager 2's Grand Tour—and the exploratory orbiters Galileo and Cassini sent to Jupiter and Saturn in its wake—was a shining light in a century plagued by almost continuous war and destruction.

The majestic achievements allowed all humanity a moment of pride and joy. For the astronomer Carl Sagan, who died in 1996, and for countless other men and women who had worked on the missions or merely observed their achievements, Voyager and all her sisters, manned and unmanned, made the end of the twentieth century the perfect time to have been alive. If humanity could leap from the sands of Kitty Hawk to the Moon's Sea of Tranquillity in a mere sixty-six years and send its robots to size up the solar system within only two decades, the promise of the twenty-first century and beyond seemed wondrous indeed.

Humans view the Moon and its Sea of Tranquillity up close for the first time, Christmas Eve, 1968. The Apollo 8 astronauts orbited for twenty hours, setting the stage for a landing seven months later.

EDITOR'S NOTE

The planetary nebula
NGC 6751, found in the
constellation Aquila.
Ultraviolet radiation
emitted from the
exposed hot stellar
core seen here in the
center of the nebula
causes the ejected
gases surrounding it to
fluoresce.

The thirty events included in the following chapters represent more than four decades of NASA's efforts and achievements in human and robotic spaceflight. These were chosen in consultation with several individuals, among them Dr. Roger D. Launius, NASA Chief Historian. Although not exhaustive, the events covered are historically significant and highlight key moments in NASA's history, offering a unique insight into the agency's past, present, and future.

Throughout the book, contributors are identified by the role they played at the time of the event. Many of these people have continued to be a part of space exploration in other roles in the private and public sectors.

Finally, for each contributor generously sharing his or her experiences within these pages, there are thousands of others whose dedication, creativity, enthusiasm, and perseverance have made NASA's accomplishments possible. *The Infinite Journey* is a tribute to each of them for allowing us to keep reaching for the stars.

Mary Kalamaras

Before Sputnik, the plan to get a man to space involved flying him to orbit in an aerodynamic, winged spacecraft and bringing him back down for a graceful reentry. But that effectively went out the window when the Soviet Union reached space with the little satellite.

Just before NASA came into being, a Joint Manned Satellite Panel was created. Two of the members of its steering committee came from the Pentagon's Advanced Research Projects Agency (ARPA) and six came from Hugh Dryden's National Advisory Committee for Aeronautics (NACA), which was about to turn into NASA. Using years of studies by scientists and engineers, and working very quickly, the committee produced a report called "Objectives and Basic Plan for the Manned Satellite Project."

The report boiled down to three objectives: launch a man into Earth orbit as soon as possible; see how he functioned there, both physically and psychologically; and get him back down safely. Safety was paramount, not only from a design standpoint, but physiologically. There was real fear that eyeballs would turn into the equivalent of soft-boiled eggs because of gravitational force and that radiation would cause cancer and other dire medical problems.

Given the fact that time was now taken to be critical, the mission would use off-the-shelf hardware as much as possible. A launcher capable of lifting an exotic rocket plane to space was decades away. What was available was the ballistic missile. By October 1958, the month NASA was born, Maxime A. Faget and his engineers decided that the rocket of choice was the Air Force's liquid-fueled Atlas ICBM (Intercontinental

When NASA selected
the first seven astro-
nauts in 1959, no one
was sure the human
body could withstand
prolonged weightless-
ness. The Mercury
flights showed that
people could work
productively in orbit.

Ballistic Missile), and that it would carry a "man-rated" capsule barely large enough to carry the man.

A Space Task Group (STG), consisting mainly of engineers, was created in November to draw up specifications for the Mercury capsule and related hardware. The McDonnell Aircraft Corporation of St. Louis, Missouri, was responsible for designing this metal cocoon—a foolproof life-support system, including special suits for space—that could safely carry a human being through the potentially deadly vacuum for the first time. This daunting challenge required close cooperation between NASA's engineers (including the soon-to-be-legendary Faget) and those from the contractors.

They came up with a blunt-nosed cone that measured roughly six feet at its widest point and a little over nine feet from its nose to the retro-rockets that would slow it as it slammed back into the atmosphere. The "crew compartment" was a claustrophobe's nightmare: a cubicle stuffed with a contoured seat and the same wraparound instrument array, consisting of dials, switches, levers, and buttons, that filled notoriously small fighter cockpits. Unlike in fighter planes, however, the astronauts would return to Earth on their backs, protected by an ablative shield that absorbed the blistering heat caused by air friction in the atmosphere. And if that weren't undignified enough, the whole contraption would then float down under a parachute until it hit the ocean in a "splashdown."

Survival school in the Nevada desert. Such training was necessary because the Mercury astronauts could have been forced down in some remote and hostile part of the globe.

Fearing a catastrophic problem with the rocket on the launchpad, Mercury's designers, like their Soviet counterparts, put an "escape tower" on top of the capsule, containing solid-fuel rockets. In the event of an emergency, the idea was to fire the rockets so they pulled the capsule off the big rocket booster, sending it high enough that the parachutes opened, carrying the astronaut safely down while the booster erupted into a fireball. Fortunately, that particular emergency never occurred.

The STG also set specifications for the astronauts. At first, the field was open to several disciplines. But the group soon concluded that if time was critical, and if the capsule necessarily had to be similar to a cockpit, then fighter pilots were the only real first choice. Accordingly, Mercury astronauts had to be under forty years old, less than five feet eleven inches tall (dictated by the size of the capsule), and in excellent physical condition, have a bachelor's degree in science or engineering or its equivalent, and be graduates of test pilot school with fifteen hundred hours or more flying time.

Right: L. Gordon Cooper is helped into his tiny *Faith 7* capsule prior to the sixth and final Mercury launch in 1963. During his thirty-four-hour flight, Cooper sent back live television pictures from space.

NASA selected 110 candidates from the files of 508 military test pilots, and reduced them to 69 for interviews. Fifty-six of those were given written exams, further reducing the number of candidates to thirty-two who went through mental and physical testing. By April 1959 the field was narrowed to seven: Captains Donald K. "Deke" Slayton, Leroy Gordon Cooper, and Virgil I. "Gus" Grissom of the U.S. Air

Force; Lieutenant Commanders Walter M. Schirra Jr. and Alan B. Shepard Jr., and Lieutenant M. Scott Carpenter of the U.S. Navy; and John H. Glenn Jr., a U.S. Marine lieutenant colonel who held a coast-to-coast speed record.

On April 12, 1961, the Soviet Union claimed another triumph when Major Yuri A. Gagarin rocketed into orbit in a Vostok capsule and became the first earthling to reach space. The race was on.

Shepard, stuffed into a Mercury capsule named *Freedom 7*, rode on a suborbital flight 303 miles down the Atlantic from Cape Canaveral on May 5. He was followed by Grissom and then by Glenn, the first American to orbit Earth. Then came Carpenter, Schirra, and finally, "Gordo" Cooper, who became the last American to go to space alone when he flew a record twenty-two orbits in *Faith 7* on May 15–16, 1963. (NASA's public affairs people winced at the name because they envisioned newspaper headlines saying "The United States Today Lost Faith" in case of an accident.) By then, America had its eyes on the Moon.

Deke Slayton was grounded because of an erratic heartbeat. But he was restored to flight status in 1972 and, sixteen years after becoming an astronaut, finally made it to space in the Apollo-Soyuz Test Project launched on July 15, 1975.

All of Mercury's objectives were achieved: it got Americans safely into space and laid the foundation for the longer Gemini missions and, finally, for the Apollo program, which won the race to the Moon.

▼

The "crew compartment" was a claustrophobe's nightmare: a cubicle stuffed with a contoured seat and the same wraparound instrument array, consisting of dials, switches, levers, and buttons, that filled notoriously small fighter cockpits.

FREEDOM 7

Stung by Yuri A. Gagarin's orbit around the world on April 12, 1961, the United States responded by sending former U.S. Navy test pilot Alan B. Shepard Jr. on a 303-mile suborbital flight downrange from Cape Canaveral on May 5.

Since Shepard was not headed to Earth orbit, an Atlas booster was not necessary, so his spacecraft was propelled by a less-powerful, modified Redstone ballistic missile. It came to life at 9:34:13 A.M. (EST) with a great roar.

"Roger, liftoff and the clock is started," Shepard reported with test pilot calmness.

"Reading you loud and clear," Deke Slayton answered from the control room.

"This is *Freedom 7*," Shepard continued. "The fuel is go. One point two g. Cabin at fourteen psi. Oxygen is go." And so was the United States. The first American in space reached an altitude of 116 miles during his fifteen-minute, twenty-two-second flight.

"No one could be briefed well enough to be completely prepared for the astonishing view I got," Shepard would later recall. "My exclamation back to Deke about the 'beautiful sight' was completely spontaneous. It was breathtaking."

Shepard was given a hero's welcome that included a reception at the White House at which President Kennedy presented him with a medal. After removing it from its velvet-lined box, Kennedy dropped it and then picked it up.

"This decoration has gone from the ground up," the quick-witted president quipped.

So, of course, had Shepard.

> The most emotional thing was Shepard's flight because you still saw these rockets blow up right and left. . . . I watched from the roadblock and I've never seen so many guys cry.
>
> **Guenter Wendt, pad leader**

Alan Shepard prepares for his fifteen-minute suborbital flight. He lost the race to be first in space by only twenty-three days.

Maxime A. Faget, assistant director of engineering and development, Manned Spacecraft Center

To be perfectly frank with you, when I was working on Project Mercury, I think a lot of my colleagues never realized that the Russians would put a man into orbit before we did, and I don't know why we were so naive, but we just [moved the program along], so it came as sort of a surprise. I think higher management was probably aware of the fact that the Russians were making progress. I can't believe we didn't know what they were doing, but it never trickled down to the level of the troops in the trenches, that's for sure.

Nevertheless, we were able to get our first flight off with Shepard just a matter of weeks after Yuri Gagarin went into orbit, and I really think that timing made it possible for the president to jump on the fact that we were in a race with the Russians and that he wanted to win. He was not a loser; he was absolutely a competitor, there's just no doubt about it. But on the other hand, if we had chosen to go to some more elaborate approaches to manned spaceflight, I suspect we'd have been very easily a year or two years late, in which case I think we'd have been out of the race right at the beginning and we would just never have come out of the blocks, as a matter of fact. We would not have gotten into the race. So in many ways, I think Mercury pretty much jump-started the whole Apollo race.

Howard Benedict, Associated Press space reporter

NASA was being very secretive about who the first man in space was going to be. We knew it was going to be John Glenn, Gus Grissom, or Al. They had been named the final three. It was amazing how NASA kept that a secret, with all the leaks that come out of that place. They would give all three of them equal time in the simulators so that even the workers didn't know. They really kept a lid on it. It wasn't until May 2 of '61 at about four in the morning— launch was scheduled for 7:30 or something like that—that I got a call from a guy who said Alan's the one who's got the suit on. I was able to call someone at NASA and verify. So we went on the wire that Alan Shepard was going to be the first American in space. He never got to the launchpad that day because of rainstorms and clouds, so they scrubbed it. But since the word was out all over the world by that time, they announced at the scrub that Alan Shepard indeed was to be the first in space. He was launched three days later.

FREEDOM 7

Alan B. Shepard Jr., commander

We had been in training for about twenty months or so, toward the end of 1960, early 1961, when we all intuitively felt that Bob [Robert R.] Gilruth had to make a decision as to who was going to make the first flight. And, when we received word that Bob wanted to see us at five o'clock in the afternoon one day in our office, we sort of felt that perhaps he had decided. There were seven of us then in one office. We had seven desks around in the hangar at Langley Field. Bob walked in, closed the door, and was very matter-of-fact as he said, "Well, you know we've got to decide who's going to make the first flight, and I don't want to pinpoint publicly at this stage one individual. Within the organization I want everyone to know that we will designate the first flight and the second flight and the backup pilot, but beyond that we won't make any public decisions. So," he said, "Shepard gets the first flight, [Virgil I. "Gus"] Grissom gets the second flight, and [John H.] Glenn is the backup for both of these two suborbital missions. Any questions?" Absolute silence. He said, "Thank you very much. Good luck," turned around, and left the room. Well, there I am looking at six faces looking at me and feeling, of course, totally elated that I had won the competition. But yet almost immediately afteward feeling sorry for my buddies, because there they were. I mean, they were trying just as hard as I was, and it was a very poignant moment because they all came over, shook my hand, and pretty soon I was the only guy left in the room.

Alan B. Shepard Jr.

I've heard it expressed a few times, and that is [that] the decision Jack [John F.] Kennedy made to go to the Moon was made after we only had fifteen minutes of total spaceflight time. A lot of people chuckle and say, "Sure!" But the fact of the matter is that, that is true. . . . Originally Louise [Shepard] and I were supposed to proceed to the Congress after the White House ceremony . . . have a reception, and then leave town. But Jack said, "No, I want you to come back to the White House, have a meeting, and let's talk about your flight." So we had the reception at the Hill, drove back, and in the Oval Office there were the heads of NASA there and the heads of the government. Jack, of course, was there. . . .

There's a picture of me sitting on the sofa, Jack is in the rocking chair, and I'm telling him how I was flying the spacecraft, and he's leaning forward listening intently to this thing. We talked about the details of the flight, specifically how man had responded and reacted to being able to work in a space environment. And toward the end of the conversation he said to the NASA people, "What are we doing next? What are our plans?" And they said, "There were a couple of guys over in a corner talking about maybe going to the Moon." He said, "I want a briefing." Just three weeks after that mission, fifteen minutes in space, Kennedy made his announcement: "Folks, we are going to the Moon, and we're going to do it within this decade." After fifteen minutes of space time! Now, you don't think he was excited? You don't think he was a space cadet? Absolutely, absolutely! People say, "Well, he made the announcement because he had problems with the Bay of Pigs, his popularity was going down." Not true! When Glenn finished his mission, Glenn, Grissom, and I flew with Jack back from West Palm [Florida] to Washington for Glenn's ceremony. The four of us sat in his cabin and we talked about what Gus had done, we talked about what John had done, we talked about what I had done. All the way back. People would come in with papers to be signed and he'd say, "Don't worry, we'll get to those when we get back to Washington." The entire flight. I tell you, he was really, really a space cadet. And it's too bad he could not have lived to see his promise.

Jerry Cunningham, USS *Lake Champlain* flight-deck crew member

At the time that Alan Shepard was launched into space I was seventeen years old. All this was new, and not too many of us knew what was actually going on. I don't really think anyone knew what to expect. The carrier's catwalks—the walkways around the flight decks—were all lined with people. Normally the flight deck was off-limits to probably 90 percent of the crew.

The captain or someone else was talking on a loudspeaker, telling everyone what was happening, and everyone was looking for this thing to come streaming down. I think the first vision I had was of spotting the helicopter coming back to the ship with the spacecraft swinging below it.

Then they brought the commander aboard. It was all hands off. Back then they were concerned about radiation contamination, and they had to check him with Geiger counters. Eventually they took him below deck.

Back in those days, to us young kids the aircraft carrier looked huge. To see this small capsule, knowing that it flew and landed in the middle of the ocean, and to recover it, was pretty amazing to all of us. There was a capsule platform aboard that had big, thick cushions on it. We had to move it around and get it strapped down so that when the helicopter brought the capsule aboard they could set it down on this platform.

Everyone was really excited and thrilled that they had been a part of it, no matter how small the part. It's not usually in the cards for an old hillbilly boy like me to have something like that happen, so of course it's been something special to me all of these years.

Robert J. Wussler, journalist and producer

Before becoming a CBS television producer who helped bring the Moon landing to millions of viewers, Wussler was a reporter covering the early "space beat."

The first flight, Shepard's, we covered out of the back of a station wagon. We had Cronkite plus maybe two reporters. There was very limited access. You needed an escort officer to go anyplace. You were not allowed on Cape Canaveral on your own at the time. It was all administered by the Air Force. . . . I remember going out to the launching pad. I was taken out by one of the NASA escorts about three in the morning. And it reminded me of a scene of a gas station on the New Jersey Turnpike in the middle of the night— hoses running all over the place, guys in hardhats. . . .

Sam T. Beddingfield, mechanical engineer

We were so busy down there with what we were doing we weren't even concerned about what the Russians were doing. I'm sure there were people in NASA who were really concerned about it, but we weren't. We were so busy just trying to make something work. It was nothing to work a seventy- or eighty-hour week. When I first started off I was the mechanical engineer, there were no others. After a few years I became the mechanical engineering division chief, because we had enough people. . . . It was a very busy time.

For Shepard's flight, I was right out there at the launchpad. I went up to where he had to get into the spacecraft and I helped him. Then at launch time, they had anybody who was close by stay in the blockhouse there, but they still wanted me and a few other guys nearby, in case we had a problem, like an emergency landing or something. We were outside so that we could get to it very quickly. I stayed there and watched the launch.

You thought of a million things that could go wrong, and when they didn't it was a big relief.

Crew members of the

U.S. Navy carrier *Lake Champlain* witness the return of Navy pilot Alan Shepard's *Freedom 7* capsule, May 5, 1961.

LIBERTY BELL 7

Virgil I. "Gus" Grissom's suborbital flight, like Shepard's, was supposed to evaluate the pilot's reaction to spaceflight and monitor his performance as an integral part of the system: how man and machine worked together in the new realm of space. It was not an auspicious relationship. The flight of MR-4 (for Mercury Redstone) was a harbinger of Grissom's untimely, tragic death almost six years later.

After being derided by other test pilots because they could not control their spacecraft, the astronauts asked for hatches they could blow open, along with other methods of manually controlling their capsules.

Grissom took off at 7:20 A.M. (EST) on the morning of July 21, 1961, when the Redstone blasted *Liberty Bell 7* off from Pad 5. It hit the water fifteen minutes later. The flight, almost an exact duplicate of Shepard's, went according to plan.

Grissom opened his helmet's faceplate, disconnected the oxygen mask, and checked in with the helicopters that were coming to pull him out of the sea. "Suddenly," he later recalled, "the hatch blew off with a dull thud." He quickly tore off his helmet and hoisted himself through the open hatchway. *Liberty Bell 7*, filled with water, soon sank out of sight. And Grissom very nearly did the same.

For the rest of his life, Grissom lived under a glare of suspicion that he himself had accidentally blown the hatch open. In July 21, 1999, almost forty years to its launch day, *Liberty Bell 7* was recovered from the floor of the Atlantic Ocean, reviving popular discussion about the events of 1961.

▼

Gus was a go-getter. He didn't talk a lot, but when he did talk, it was worth listening to. And although we're all cut from the same cloth, he had his strengths that were unequaled by any of the other guys.

M. Scott Carpenter, Mercury Seven astronaut

Gus Grissom talks with backup pilot John Glenn before his *Liberty Bell 7* launch. Grissom's flight was a repeat of Shepard's suborbital shot.

Virgil I. "Gus" Grissom, commander

In his post-flight report, the second American in space recounts his actions following his spacecraft's successful landing in the Atlantic Ocean.

I felt that I was in good condition at this point and started to prepare myself for egress [exit]. I had previously opened the faceplate and had disconnected the visor seal hose while descending on the main parachute. The next moves, in order, were to disconnect the oxygen outlet hose at the helmet; unfasten the helmet from the suit; release the chest strap; release the lap belt and shoulder harness; release the knee straps; disconnect the biomedical sensors; and roll up the neck dam. The neck dam is a rubber diaphragm that is fastened on the exterior of the suit below the helmet-attaching ring. After the helmet is disconnected, the neck dam is rolled around the ring and up around the neck, similar to a turtleneck sweater. This left me connected to the spacecraft at two points: the oxygen inlet hose, which I needed for cooling, and the helmet communications lead.

At this time, I turned my attention to the door. First, I released the restraining wires at both ends and tossed them toward my feet. Then I removed the knife from the door and placed it in the survival pack. The next task was to remove the cover and safety pin from the hatch detonator. I felt at this time that

everything had gone nearly perfectly and that I would go ahead and mark the switch position chart as had been requested.

After about three or four minutes, I instructed the helicopter to come on in and hook onto the spacecraft and confirmed the egress procedures with him. I unhooked my oxygen inlet hose and was lying on the couch, waiting for the helicopter's call to blow the hatch. I was lying flat on my back at this time, and I had turned my attention to the knife in the survival pack, wondering if there might be some way I could carry it out with me as a souvenir. I heard the hatch blow—the noise was a dull thud—and looked up to see blue sky out the hatch and water start to spill over the doorsill. Just a few minutes before, I had gone over the egress procedures in my mind, and I reacted instinctively. I lifted the helmet from my head and dropped it, reached for the right side of the instrument panel, and pulled myself through the hatch.

Jim Lewis, U.S. Marine Corps, primary recovery helicopter pilot

I had made contact with Gus and told him we were in the area. A couple of times he asked for more time to finish his shutdown checklist and closeout checklist for the instrumentation. That was OK with us. We had plenty of fuel and the capsule was riding well in the water.

We'd made the turn and were on final approach to the capsule when the hatch blew, which was a huge surprise to all of us. Gus had been floating in the capsule and looking out of the window at the Atlantic Ocean. The water outside was at his eye level, halfway up the hatch and window. He certainly could not have blown it, simply because he saw the water and also because he knew the procedure: We had to come in, cut the capsule's antenna, hook on, tell him we'd done that, then lift the capsule out of the water so that the hatch would be clear. None of that had happened yet.

The hatch did blow, for whatever reason, before we finished our approach, and the capsule sank awfully fast. Gus came out immediately. I wasn't worried about him because we had practiced with astronauts and I knew they could float in the water quite easily. I was intent on trying to recover the capsule at that point. We got in and cut the antenna as it was sinking. By the time my copilot, John Reinhard, finished that, the recovery loop at the top of the capsule was just disappearing below the surface of the water. John managed to hook on as it went out of sight. Before we ever tried to lift the capsule up, we got an engine warning light in the cockpit. We knew from all our manuals and from our training that if we ever got one of those lights, the engine would continue to run for about five more minutes before stopping or freezing. I told John, who was in the process of lowering the sling to pick up Gus, to not lower the sling any more because we had a sick bird. I didn't want Gus in the plane in the event we didn't make it back to the carrier. Ditching a helicopter in water is not an easy thing because when you go into the water, the rotor is still turning over your head. We had never practiced that particular scenario with the astronauts. I immediately called the backup helicopter and told them to come in and pick up Gus while I started dragging the capsule out of the way.

Virgil I. "Gus" Grissom, commander

During their struggle to recover the capsule, the helicopter crew is unaware that Grissom is fighting to stay above water.

The helicopter pulled up and away from me with the spacecraft and I saw the personal sling start down; then the sling was pulled back into the helicopter and it started to move away from me. At this time, I knew a second helicopter had been assigned to pick me up, so I started to swim away from the primary helicopter. I apparently got caught in the rotorwash between the two helicopters because I could not get close to the second helicopter, even though I could see the copilot in the door with a horse collar swinging in the water. . . . When I first got into the water, I was floating quite high up; I would say my armpits were just about at the water level. But the neck dam was not up tight and I had forgotten to lock the oxygen inlet port; so the air was gradually seeping out of my suit. . . . I was going underwater quite often. The mild swells we were having were breaking over my head and I was swallowing some salt water. As I reached the horse collar, I slipped into it and I knew I had it on backward; but I gave the "up" signal and held on because I knew that I wasn't likely to slip out of the sling. As soon as I got into the helicopter, my first thought was to get a life preserver so that if anything happened to the helicopter, I wouldn't have another ordeal in the water.

A rescue helicopter struggles to pull *Liberty Bell 7* from the sea moments before the waterlogged spacecraft sinks beneath the waves.

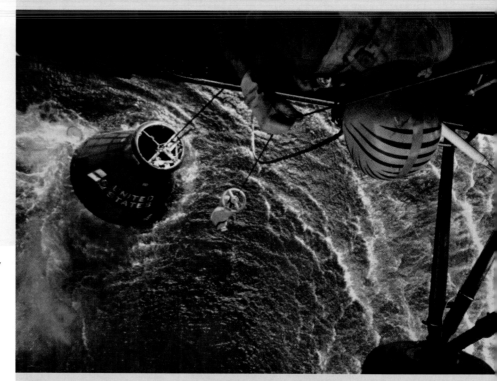

In 1999, an expedition team recovered *Liberty Bell 7* from the Atlantic Ocean floor nearly four decades after it was lost. The restored capsule will reside at the Kansas Cosmosphere and Space Center.

I tried to lift the capsule as I was dragging it. If there had been more time it might have been possible to have gotten it out. After five minutes, the gauges started going in the wrong direction and, expecting imminent engine failure, I released the capsule, declared an emergency, and headed back to the carrier. . . . When the hatch blew, the capsule sank so fast that it was really fortunate that we were able to hook on at all. I never felt a sense of failure. Nobody likes to fly a plane back with less than a perfect mission, but having been dealt what we were dealt, I thought that it was pretty close to a perfect mission. When we got the capsule back thirty-eight years later, that finished it. Seeing the capsule break the surface at two A.M.—after being awake for twenty-four hours and waiting with some uncertainty—provided one of the most spectacular sights and feelings of my life.

Max Ary, president and CEO, Kansas Cosmosphere and Space Center
Ary, along with expedition leader, Curt Newport, was involved in the dramatic recovery of the lost Liberty Bell 7 *capsule on July 21, 1999.*

My professional interest in *Liberty Bell* picked up in the late 1970s. We were just starting to assemble the collection for the Cosmosphere and we had this big checklist of things that we needed to do to have a definitive collection. Of course you needed to have a Mercury, a Gemini, and an Apollo spacecraft. We'd acquired an Apollo spacecraft; I pretty much knew where I was going to be able to find a Gemini, but the one big hole was going to be Mercury because they were so rare. They built the fewest number of those spacecraft, and all of them at that point were in museums. I didn't think there was any chance of getting one away. The only Mercury that I knew of that no one had their hands on was sitting at the bottom of the ocean. That was the *Liberty Bell*.

With the exception of the birth of my children, seeing the *Liberty Bell* come up from the ocean was one of the most surreal events I have ever experienced. I'd spent almost twenty years thinking about that spacecraft. I'd constantly have these images in my mind of what it looked like sitting on the bottom. That night—especially because it was at night and we had lights down on the water — was kind of eerie. When that capsule came up out of the water, it didn't seem real. I still get goosebumps.

▼

I think Gus would be really pleased to see *Liberty Bell* back home—the only craft he didn't bring back. I think about Gus and all of the things that he did and all that he accomplished. He was a tremendous individual: a super engineer; a skilled, experienced test pilot; actually the astronaut's astronaut. What a guy. . . . I know he'd be pleased, very pleased to see it back.

Lowell Grissom, brother to Gus Grissom

FRIENDSHIP 7

J ohn H. Glenn Jr. roared into the sky over Cape Canaveral on the morning of February 20, 1962, on top of an Atlas D, on his way to becoming the first American to orbit Earth.

Technically, the flight of *Friendship 7* was originally supposed to evaluate the man-machine system. But with Yuri Gagarin having gone to orbit first the previous April, and Gherman Titov having flown sixteen orbits in Vostok 2 only four months later, the political pressure to catch up was phenomenal.

Glenn enjoyed a spectacular view of the Canary Islands and the West African coast, and then sailed into the night over the Indian Ocean and Australia, and on to a brilliant dawn over the Pacific. It was then that he noticed "fireflies" racing past his window. Later determined to be fluid jettisoned by *Friendship 7*'s environmental system and turned into harmless ice particles, these "fireflies" would not turn out to be a problem for Glenn.

Instead, both the astronaut and his ground controllers thought the heat shield might have been tearing off. If true, *Friendship 7* would have disintegrated in a fireball. He was therefore ordered down after completing only three of seven planned orbits.

Glenn dropped into the Atlantic after nearly five hours in space and was given a tumultuous welcome when he got home by hero-hungry Americans, including a ticker-tape parade in New York reminiscent of the one for Charles Lindbergh after his solo flight across the Atlantic in 1927.

An onboard camera catches John Glenn inside the Mercury spacecraft during his five-hour flight. He called the view "beautiful" and found weightlessness to be a pleasant experience.

M. Scott Carpenter, backup astronaut

I was John's backup, and part of that job was to be in the blockhouse during the count, and that's where I was. And I was taking care of all of the communications from the launch people and the launch complex to John. And I was, so I was told, the only one who would be able to communicate with John in that period from T-minus eighteen seconds to liftoff. That's when it occurred to me that this fellow named John Glenn, in order to have a successful flight, was going to have to put under his belt more speed than we had ever given a human before. Speed was the essence. If he could get the speed and if it were in the right direction, he had orbital flight licked. You know, "Godspeed" is something you hear all the time; but speed was very, very important to John. And it just came to me, "Godspeed, John Glenn"; and I think the fact that his name is two short syllables made it ring a little better. But anyway, somewhere in the count between ten and zero I said, "Godspeed, John Glenn." And it was a salute to him, but there was a feeling, I think, in me at the time that it could be viewed as a plea to whatever Higher Power to, you know, make this flight a success. And I would suggest that nobody can tell me that that plea didn't work, because the flight did.

With mechanical engineer Sam Beddingfield (left) looking on, Glenn gets ready for his historic flight, February 20, 1962. The name *Friendship* was chosen by the astronaut's family.

FRIENDSHIP 7

John D. Hodge, assistant chief for flight control, Manned Spacecraft Center

Teamwork between ground crew and astronaut contributed to the Mercury program's success. In the case of Friendship 7's heat-shield problem, the ability to make real-time decisions was tested. Hodge played a major role in establishing mission rules and flight control tasks early on.

Because everything was sort of unknown and there were so many things that could go wrong, we had to try to anticipate anything that could go wrong. We would dream up a problem and then find a solution to it and write that solution down. As a result, you developed a whole process of thinking on how to make a decision should something go wrong. You have to remember that if something goes wrong while the spacecraft is over a tracking ship, say, in the Sea of China, that ship will see the spacecraft for, at the most, eight or nine minutes. If something goes wrong in that time, you have to be able to work a process out whereby you can make a decision in that very short time. That thinking process that we wrote down eventually became mission rules. . . .

When we first started, we did what we called "paper simulations." We had a hangar down at Langley [Air Force Base] that was separated off with blue hanging sheets, and each one of these little "rooms" represented a remote station. There was Bermuda remote station or the RKV [the *Rose Knot* tracking ship] or Mexico or California, or something like that. The people who were running the simulation would send notes that said, for example, "liftoff," to teletypes at every one of the stations. Then they would send around a message that said such and such a thing had happened. We had dozens of secretaries rushing around with these notes to everybody, trying to get this thing going. It was very, very primitive in the beginning. We did that for close to a year. When we finally got the network finished, we sent the people out several weeks before a mission. We would run a mission as though it were real because everyone was sitting at their consoles with data coming in. We would send teletypes out with all of the information. It worked extremely well.

John H. Glenn Jr., commander

There are many things that are so impressive, it's almost impossible to try and describe the sensations that I had during the flight. I think the thing that stands out more particularly than anything else is the reentry during the fireball. I left the shutters open specifically so I could watch it. It got to a brilliant orange color; it was never too blinding. The retropack was still aboard and shortly after reentry began, it started to break up in big chunks. One of the straps came off and came around across the window. There were large flaming pieces of the retropack—I assume that's what they were—that broke off and came tumbling around the sides of the capsule. I could see them going on back behind me then and making little smoke trails. I could also see a long trail of what probably was ablation material ending in a small bright spot similar to that in the pictures out of the window taken during the MA-5 [Mercury-Atlas] flight. I saw the same spot back there, and I could see it move back and forth as the capsule oscillated slightly. Yes, I think the reentry was probably the most impressive part of the flight.

Alan Shepard, capsule communicator (CapCom) for the *Friendship 7* mission, gives a thumbs-up as Glenn's Atlas rocket climbs into orbit. The orbital flight achieved Mercury's primary goal.

John H. Glenn Jr.

The main purpose of my flight was to find out what reaction the human body had to extended weightlessness. Some of the doctors . . . thought that my vision might change during flight, because when the eye no longer had to be supported by the structure under the eye, it might gradually change shape, and if it did, you might get horribly myopic or something where you couldn't see properly. So, on the instrument panel—and you can still see this up in the Smithsonian [Institution] on my spacecraft—is a little Snellen chart like the eye chart they use in doctors' offices, miniaturized for the distance from my eyes to the panel. I was to read the smallest line I could read every twenty minutes all during flight. . . .

Some of the doctors at that time felt that when you're weightless . . . you might get uncontrollable nausea and vertigo and [would] not be able to make even an emergency reentry because you'd have such nystagmus [oscillation of the eyeballs] that you wouldn't be able to see the instrument panel properly. They were very concerned about this, enough so that they had not only [motion sickness] pills that I was to take if I started feeling the least bit woozy . . . they also had that stuff in a solution. It was in a pocket on my leg in a special syringe . . . if I needed it I could take it out, take the safety catch off, hit my leg, and a spring would be released that drove the needle through the suit into my leg and injected the serum. . . . That was to be a get-me-down type operation, if I was getting so out of control that I felt I had to make an emergency reentry. . . .

They didn't know whether you could swallow properly or not. I wasn't going to be up long enough that I really had to have a meal or two meals or anything, but they wanted me to take material along to swallow, which I did. They wanted to know if there was any change of feeling, in my fingers or anything like that, any tendency toward any sickness, whether it was induced from the inner ear or wherever. This was more to find out the body's reaction to flight so we'd know whether we had to make any adaptation before we could go on to longer flights or to the flights that would later build up to go to the Moon.

L. Gordon Cooper Jr., astronaut

We had the tracking site north of Perth, Australia. And, of course, we had a lot of delays in getting John launched, so I was there for quite a period of time. We had a joint team there composed of both the Australians and Americans. It was a good bunch. . . . They did a great job on the tracking side.

There was a question about what could an astronaut really see on nightside, in the dark; what could he see in the way of lights? So the people of Perth, including the mayor of Perth, decided that they would bring up all the lights they could. And British Petroleum turned up some big oil-fired torches that they had out south of Perth that put out umpteen thousand BTUs [British Thermal Units] per torch; they turned all those on. And everybody in Perth brought up their lights to see if John could see the city. Of course, he could see almost each and every light, but it was a well lit-up area. . . . It really captured the imaginations of people, I think.

In space, the Sun rises every ninety minutes. Glenn radioed back to Earth: "That sure was a short day."

Sam T. Beddingfield, mechanical engineer

I was a twenty-five-year-old engineer in charge of the escape system for Project Mercury. I remember we were launching John Glenn when we noticed a broken screw in one of the Mercury capsule's heat shields. It would have taken several hours to repair, so I made a decision. We plugged the screw hole with RTV-90 [a rubbery sealing compound], and we went ahead and flew. We didn't even tell anybody. You couldn't do that today. Now it would take a committee a week to make a decision like that. In those days, young engineers were in charge of entire systems. There weren't layers of management on top to challenge every decision.

THE GEMINI TWINS

Exploration is by definition the extension of missions: ever farther, ever more complex. Gemini's purpose was to form a link between Mercury's achievement of sending individuals on short preliminary orbital missions around Earth and Apollo's goal of carrying three men to a landing on the Moon.

Since the voyage to the Moon would require maneuvering on the way to it and, once there, more maneuvering down to the lunar surface and then back up to the Apollo command and service modules, moving the spacecraft was one of Gemini's primary goals. So, too, were rendezvousing, docking, space "walks" (extravehicular activity, or EVA, in space jargon), and longer-duration missions to test the astronauts both physically and psychologically.

On December 7, 1961, NASA associate administrator Robert C. Seamans approved what was then called Mercury Mark II. Less than a month later, in recognition of the fact that its mission would be to fly two astronauts, the program's name was changed to Gemini after Castor and Pollux, the bright, near-twin stars in the Gemini constellation.

The plan called for Gemini capsules to be launched by modified two-stage U.S. Air Force Titan II ballistic missiles, which were more powerful than the Atlases used to boost Mercury capsules. This was necessary because the new spacecraft weighed more than twice

Gemini 7 as photographed from Gemini 6 during the first space rendezvous, December 1965. The meeting of two spacecraft in orbit—Gemini's main accomplishment—was a critical step in reaching the Moon.

Jim McDivitt (left) and
Ed White training in a
mock-up for their
Gemini 4 flight. As the
first of the Gemini pro-
gram's longer missions,
this flight pressed new
demands upon ground
control, including the
need to move into a
three-shift operations
plan.

▼

Gemini added 970 hours
of flight time to the U.S.
program and proved that
astronauts could func-
tion smoothly together
in space, work outside
the spacecraft, and not
suffer serious physiolog-
ical damage from zero
gravity and cosmic rays.

as much as Mercury, including a two-part adapter module that connected it to the Titan and which was jettisoned before reentry. Since rendezvous and docking were the most important of the program's missions, NASA also acquired the U.S. Air Force's reliable Agena unmanned spacecraft to use as the target to which the astronauts would connect. Typically at the dawn of the space age, there were problems galore. The Titan booster, itself still in the development stage, had a spate of nasty habits: It oscillated longitudinally, bouncing like a pogo stick, and had a fuel combustion problem in the second stage and other flaws, all of which the air force and its prime contractor, the Martin Company, worked frantically to correct. By the spring of 1964, Titan was ready.

Meanwhile, the spacecraft's fuel cell leaked, and the Agena target craft had its own daunting array of difficulties, many of which continued throughout the operational phase of the program. And landing on land, as originally planned, had to be abandoned because of design problems with the paraglider. While Gemini was not exactly soaring, its cost was: It climbed from $350 million to more than $1 billion.

Following two unmanned test flights in April 1964 and January 1965, Gus Grissom and John W. Young rode Gemini 3 to orbit on March 23, 1965. Grissom, ever mindful of *Liberty Bell 7*'s loss after its hatch blew almost four years earlier, showed his sardonic sense of humor by naming his new spacecraft *Molly Brown* after an unsinkable heroine in a Broadway show. The flight demonstrated that orbital maneuvers could be done and that the astronauts themselves could manually reenter the atmosphere (refuting test pilots who derisively called them "Spam in a can" because Mercury flew by computer).

NASA's first EVA came on June 3 when James A. McDivitt and Edward H. White went up in Gemini 4. White, secured to the spacecraft by a tether (the "ultimate leash," as the astronauts called it), flew in formation with it at roughly seventeen thousand miles an hour for twenty minutes before being pulled in despite his reluctance. Three months earlier, cosmonaut Aleksei Leonov had scored yet another Soviet first by doing the same thing outside of *Voskhod 2*, nearly killing himself in the process.

"Gordo" Cooper and Charles "Pete" Conrad Jr. extended time in orbit to eight days when they went up in Gemini 5 on August 21, 1965. But there was far more to the mission than setting an endurance record. One key reason for the flight was to evaluate the rendezvous guidance and navigation system and practice with an Agena. But another glitch with a fuel cell prevented an actual docking, so Cooper and Conrad had to practice with a "phantom" target.

No phantom was needed on December 15. Eleven days earlier, Frank Borman and James A. Lovell swung into orbit on what would become the longest Gemini flight, Gemini 7. On the fifteenth, Walter M. Schirra and Thomas P.

Stafford left Canaveral onboard Gemini 6. Roughly six hours after launch, they pulled to within a foot of Borman and Lovell. All four sailed across the sky in very tight, though not yet connected, formation.

Docking would not take place until the next mission, even as the Apollo program was moving into high gear. On March 16, 1966, Neil A. Armstrong and David R. Scott inched Gemini 8 into the end of an Agena target spacecraft. It was a historic moment for the U.S. space program on the way to the Moon, but a short one because of stability problems that caused the joined spacecraft to pitch, roll, and spin at one revolution per second until Armstrong could regain control.

If anybody needed a reminder that space was a perniciously difficult domain, it happened when Stafford and Eugene A. Cernan went up in Gemini 9 on June 3. One malfunction after another, including problems with two separate target vehicles, dogged the mission. The best the hapless astronauts could do was simulate a lunar module rendezvous and go home.

As is generally the case in space, the glitches were turned to success. The last three Gemini missions, beginning on July 18, 1966, and ending with the launch of Gemini 12 on November 11, had successful dockings. And all three had successful EVAs.

Gemini added 970 hours of flight time to the U.S. program and proved that astronauts could function smoothly together in space, work outside the spacecraft, and not suffer serious physiological damage from zero gravity and cosmic rays. There were fifty-two experiments completed on the ten missions, and starting with Gemini 4, flight control operations moved from Cape Canaveral to its present-day Houston location. NASA was closing in on the Moon.

The Gemini 8 astronauts view the Gemini Agena Target Vehicle in orbit, March 1966. Neil Armstrong and Dave Scott successfully docked with the Agena, but had to abort the mission when their vehicle went into a spin.

GEMINI 4

On the morning of June 3, 1965, James A. McDivitt and Edward H. White blasted off from Cape Canaveral's Pad 19, undertaking the most daring space mission up to that time. The astronauts stayed in space for a record sixty-two orbits completed in almost ninety-eight hours; however, a fuel shortage and a computer glitch prevented them from docking with a Gemini Agena Target Vehicle (GATV).

The highlight of the mission came when White climbed out of the spacecraft while it was flying between Hawaii and Mexico at three hundred miles a minute and did America's first "umbilical" EVA (extravehicular activity). Locked onto a twenty-five-foot, gold-plated tether and flying in close formation with the spacecraft, White sailed over Earth as he joyously maneuvered himself using a handheld gas gun that squirted oxygen.

If there was one problem with an American's first walk in space, it was that the American was so excited and exhilarated by what he was doing that he didn't want to get back into the spacecraft. White was supposed to climb back into Gemini 4 while it was still in sunlight. But he wouldn't do it. For more than twenty minutes, he used his gas gun to somersault, pirouette, float on his back, and grin at McDivitt like a swimmer under water.

"They want you to get back in now," McDivitt told White.

"This is fun," the exuberant space walker replied. "I don't want to come back in, but I'm coming." And he did.

Before the Gemini 4 flight, Chris Kraft [Gemini flight operations director] and I were getting phone calls clear up through the night before launch saying we were going to kill these guys; that they weren't going to survive and would not be able to get out of the spacecraft.

Charles A. Berry, NASA director of medical operations and chief flight surgeon

McDivitt (foreground) and White inside their spacecraft. Gemini 4 marked the first time that actual mission control was transferred from Cape Canaveral (aka Cape Kennedy 1963–1973) in Florida to ground control operations in Houston, Texas.

GEMINI 4

Gemini 4's McDivitt

and White practice

egress, or exit, proce-

dures with their space-

craft while in a water

tank in Texas. Like the

Mercury astronauts

before them, the

Gemini crews splashed

down in the ocean.

Gemini 4's McDivitt and White practice egress, or exit, procedures with their spacecraft while in a water tank in Texas. Like the Mercury astronauts before them, the Gemini crews splashed down in the ocean.

Charles A. Berry, NASA director of medical operations and chief flight surgeon

We had some secret film that had come to us through the CIA of cosmonaut Aleksei Leonov's space walk three months earlier. I can't tell you how many times we watched that damn film. We couldn't figure out what he was doing when he kept hitting his leg with one hand. We thought that maybe he was trying to relieve the pressure in his spacesuit. Only recently, I met Leonov in London and he told me that it was a camera the KGB had put on his suit. The camera wasn't working, so he kept hitting it to try to make the thing work. But watching this film at the time, we could tell that he was having some trouble, so our worry level was up some.

Another thing that concerned us greatly was getting back into the spacecraft. It was a real difficult task to get back down into the seat, fasten in, and then be able to close the spacecraft hatch. As a matter of fact, Ed White was probably in the best shape of any astronaut we had. Had he not been, I think we probably would not have been able to get that hatch closed.

James A. McDivitt, commander

When Ed went to open up the hatch, it wouldn't open. I said, "Oh my God," you know, "it's not opening!" We chatted about that for a minute or two and I said, "Well, I think I can get it closed if it won't close." But I wasn't too sure about it. I thought I could. But remember, I would be pressurized . . . leaning over the top of the thing with a screwdriver. I'd be there pressurized, in the dark. So anyway, we elected to go ahead and open it up.

We didn't bother telling the ground about that. I mean, there was nothing they could do. They would've said no, I'm sure. Anyway, we went ahead and opened it up and Ed went out and did his thing. That was one of the reasons I was kind of anxious to have him get back inside the spacecraft because I'd like to do this in the daylight. . . . But by the time he got back in it was dark. When we went to close the hatch, it wouldn't close. It wouldn't lock. And so in the dark I was fiddling around over on the side where I couldn't see anything, trying to get my glove down in this little slot to push the gears together. Finally, we got that done and got it latched.

And the next part of the plan was to get Ed to repressurize the spacecraft, get all this junk off him, open up the hatch, and throw all of it out. And there was no way I was going to do that! We carried all that stuff through the rest of the flight.

James A. Lovell Jr., backup crew member

EVA was perhaps the hardest thing to do, because people forgot about Newton's third law of motion. [For every action there is an equal and opposite reaction.] The spacecraft was actually repelling astronauts when they tried to work around it without having the proper handhold.

I was White's backup for Gemini 4. In those early days we had little hand-held thrusters that didn't work very well. The idea was that you fire the thing, aiming it where you want to go. But in reality, you could never figure out where your center of gravity was, so you'd always end up tumbling or going the wrong direction or something like that.

Ed was not outside long enough to really understand the problems confronting actually going outside and working. We didn't do anything of that nature through Gemini 9. Then with Gemini 9, Gene Cernan was to go around to the back of the spacecraft, where we had mounted an astronaut maneuvering unit. He was to get in that and then actually try to maneuver without attachments at all. Very fortunately, that did not occur. When he tried to do it, he got out and started to fight the zero gravity. His heart rate started to go up, he perspired, and his visor fogged over. He had a devil of a time trying to get that backpack on his back, and finally he gave up. So we realized that working outside was a little difficult.

So by the time Gemini 12 came along, we realized that we had to learn what to do in extravehicular activity. So for the very first time in space history, we used water as the medium to simulate zero gravity. We rented a swimming pool up in Baltimore, and we sank a crude mockup of the spacecraft in this pool. Buzz Aldrin was my pilot on that flight. We put Buzz in a spacesuit and weighted him down to make him neutrally buoyant. I sat on the edge of the pool as if I was in the commander's seat and started instructing him on what to do. Then we learned how to make various handholds and toeholds and to work in zero gravity, using it as an asset rather than a liability. That was the beginning of the use of water tanks for extravehicular activity.

But White just floated out there. He didn't want to come in. McDivitt practically forced him in, it was such a beautiful view out there.

▼

One Saturday morning we were sitting having breakfast and I said, "Kids, I'm going to tell you something really important. You know that Dad's an astronaut and the astronauts fly in space. I just want to let you know that I'm going to fly in space soon." And my older boy, Mike, who was probably seven or eight, says, "Oh yeah, Dad, I heard that at school." And then my daughter Ann said, "Oh yeah, Dad, I heard that at school, too." And my son Patrick said, "Dad, there's a fly in the milk bottle."

James A. McDivitt

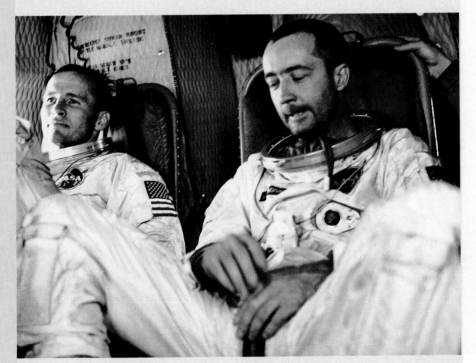

Tired but happy following a four-day flight, the just-returned Gemini 4 crew relax onboard their rescue helicopter

GEMINI 4

Mission commentary reveals a reluctant White called in from his EVA, much like a young boy called in from a playground.

McDivitt: . . . We're coming over the west there and they want you to come back in now.

CapCom [Gus Grissom]: Roger. We've been trying to talk to you for a while here.

White: This is fun.

McDivitt: Well, back in. Come on.

White: . . . to come back to you, but I'm coming.

McDivitt: You still have three and a half or four days to go, buddy.

White: I'm coming.

McDivitt: OK.

CapCom: You've got about four minutes to Bermuda LOS [loss of signal]

CapCom: Gemini 4, Houston CapCom.

White: I'm trying to. . . .

McDivitt: OK. OK. Don't wear yourself out now. Just come on in.

McDivitt: How are you doing there?

White: The spacecraft really looks like it's . . . because whenever a piece of dirt or something goes by it always heads right for that door and goes on out.

McDivitt: OK. Whoops, take it easy now.

White: OK. I'm on top of it right now.

McDivitt: OK, you're right on top. Come on in then.

White: The handhold on that spacecraft is fantastic. You can really . . . Aren't you going to hold my hand?

McDivitt: No, come on in the. . . .

McDivitt: Ed, come on in here.

White: All right.

After a few more minutes, White finally resigns himself to entering the capsule.

McDivitt: Come on. Let's get back in here before it gets dark.

White: It's the saddest moment of my life.

McDivitt: Well, you're going to find it sadder when we have to come down with this whole thing.

White: I'm coming.

McDivitt: OK. Come on now.

CapCom: Gemini 4, Houston CapCom.

McDivitt: Let's see you get those hand dogs fixed now.

CapCom: Gemini 4, Houston CapCom.

McDivitt: I'm just putting all this stuff down here . . . I'm going on in a minute. No time to talk now—I'm pulling in his air hose. OK. Any messages for us, Houston?

Flight Director [Ken Kleinknecht]: Yeah, get back in.

**Listening to congratulations from President Lyndon Johnson onboard the aircraft carrier U.S.S. *Wasp*. White's brief space walk was a smashing success for the young space program.

Opposite

Tethered to his ship, White holds a "zip gun" for maneuvering in space. Reentering the capsule, he said, was "the saddest moment of my life."

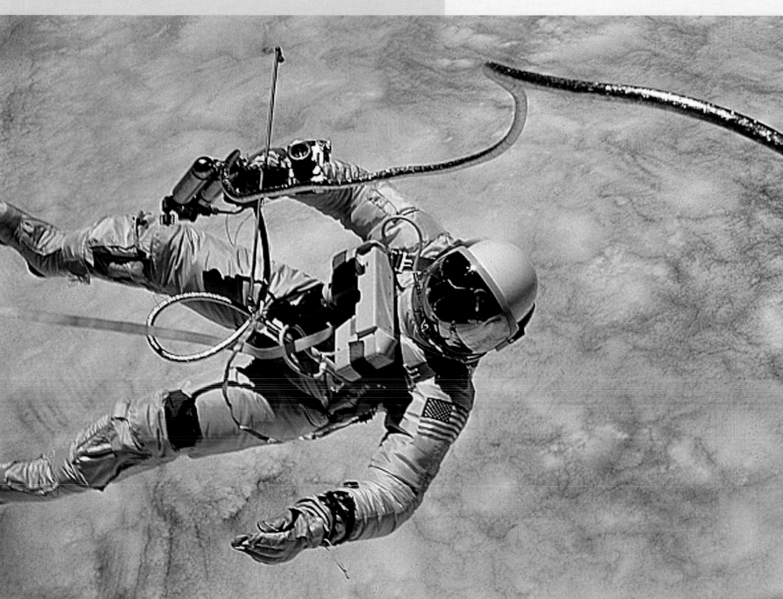

Richard W. Underwood, supervisory aerospace technologist, space photography

The pictures that McDivitt took of White became world famous. They were on magazine covers around the world. Everybody felt that it was a big publicity stunt for NASA, but if you look at the pictures, White has a camera. What White was doing was photographing the outside of the Gemini. The problem was that the Mercurys and the Geminis, and the early test vehicles before that, would become severely damaged in flight. Some engineers said, "Well, that happens on the way up. When you put a capsule on a giant rocket and pump up the g forces, things break. . . ." Other engineers were saying, "You're all wrong. It happens on the way back. The capsule becomes a flying fireball under great pressure, and things break. . . ." So at a meeting one day we said, "Why don't we send someone outside with a camera to photograph the capsule? If the damage is there, it happened on the way up. And if the damage isn't there, it happened on the way back." So White took thirty-nine photographs of the outside of the spacecraft and McDivitt took twenty-four pictures of him. It turned out that some of the damage happened on the way up and some of it happened on the way back. So the configurations were changed and the spacecraft was built better. A lot of things learned were incorporated right on down to space shuttles. . . . Ed White's pictures changed spacecraft design in certain areas, and the pictures that McDivitt took of him to try to record it are some of the most famous pictures in history.

I thought, What do you say to 194 million people when you're looking down at them from space? Then the solution became very obvious to me. . . . They don't want me to talk to them. They want to hear what we're doing up here. . . . So what you heard were two test pilots conducting their mission in the best manner possible.

Edward H. White, pilot

GEMINI 8

Amerca's first docking in space also provided the space program's first real emergency in orbit, a heart-stopping reminder that what would amount to a minor problem on Earth can become catastrophic at seventeen thousand miles an hour. Neil A. Armstrong and David R. Scott roared away from Cape Canaveral (also known then as Cape Kennedy) at 11:41 A.M. (EST) on March 16, 1966. After meticulous maneuvering, Armstrong guided Gemini 8's nose into the Gemini Agena Target Vehicle's docking collar six hours and thirty-four minutes later.

"Flight," Armstrong advised Houston, "we are docked." It was a historic moment. But the success was short-lived.

While passing over China (and temporarily out of radio contact with ground stations), Scott noticed that what had been a smooth flight a minute before was suddenly turning violent. A problem with their spacecraft's control system turned on a maneuvering thruster, which caused Gemini 8 and the twenty-six-foot satellite to which it was locked to start rolling and yawing. It spun at one revolution a second and fishtailed violently from side to side. As the seconds passed, the spacecraft, loaded with volatile fuel, turned into a twisting, tumbling bomb.

Armstrong fought with the controls. Just as he seemed to be regaining control, the spacecraft went totally crazy. "The rates of tumbling increased to a point where we felt the integrity of the combination. . . . was in jeopardy," he would later recall. The veteran X-15 rocket pilot pulled away from the Agena and used Gemini's reentry thrusters to make a forced landing in the Pacific.

▼

The liftoff, launching, rendezvous, and docking were really tremendous. We were really looking forward to the whole mission.

David R. Scott, pilot

Neil Armstrong (foreground) and Dave Scott head up the ramp to their waiting Titan II launch vehicle, March 1966. A few hours later, they would find themselves in the first space emergency.

GEMINI 8

John D. Hodge, flight director

NASA's first rendezvous and docking mission provided an unexpected turn of events for Hodge, who was debuting as mission flight director.

By Gemini 8, we had the control center in Houston. The Agena took off first, and the idea was that the Gemini would take off on the next orbit and rendezvous. It was during the third orbit that Armstrong and Scott were going to rendezvous, sort of over Madagascar. We had a remote station there and that was the last place that we could talk to them. They did rendezvous and everyone was elated. They left Madagascar knowing that we would see them in about twenty-five minutes or so. After the third orbit, you tend to go "off the network," meaning that you wouldn't see the spacecraft for long periods of time.

Well, the next thing we heard was that the spacecraft was over the Sea of China. We had a tracking ship there called the RKV (*Rose Knot* Victor), and we got a message from it saying that the spacecraft was spinning and had used up all of its control fuel. Neil Armstrong was straightening it out. To do that, he used half of his reentry control fuel.

The crew separated from the Agena, but the problem turned out to be in the spacecraft; it was the last thing they expected. You always worry about the newest thing, not the thing that you have already tested. Neil Armstrong and Dave Scott did a fantastic job in getting it under control and retained the second half of their fuel for reentry. We had that information in Houston as it was happening.

The next station for spacecraft contact was Hawaii, which was about twenty minutes later, so in that twenty minutes we had to decide: Did we really know what was wrong? Should we come down immediately? Was it safe to come down? And, if so, where would we come down? Since we were going off the network, we only had that one period of eight or nine minutes while the crew was over Hawaii to decide on what to do. So we all got together and talked about it, and we talked to the recovery people and made the decision that the only thing to do was to come in on the next orbit. Unfortunately, coming in on the next orbit meant firing the retrorockets over Tibet and landing in the Sea of China, where the recovery people had a destroyer. The question was, how accurate would all that be? It was the tradeoff you had to make, or wait another half a day until you had more information and hope that the spacecraft would not spring another leak. So we decided to come in, and it was nail-biting time. But as it turned out, they came down quite close to the ship, and everything was fine. But it was nail-biting, no question about it.

Neil A. Armstrong, commander

Someone's Law guarantees that bad things always happen at the most inconvenient times. It was certainly true for Gemini 8. We were attached to another spacecraft, Agena; we were on the nightside of Earth with little visual reference; and we were out of range of any of our Earth-based tracking stations and could not ask for advice from our colleagues in Mission Control or other monitoring groups. We would have to deal with the problem ourselves.

Dave Scott actually noticed the problem first. We were slowly rolling. We both suspected the Agena spacecraft. We normally could hear the Gemini's attitude thrusters when they fired, and we weren't hearing anything. Dave controlled the Agena from a panel on his side of the cockpit. He was trying every trick he knew to bring the Agena under control, to no avail.

With roll rate increasing to uncomfortable levels, we reluctantly concluded that we must undock and solve the mystery in a simpler, one-spacecraft configuration. So we bid adieu to Agena and relatively quickly thereafter chose a course of action that would return us to a stable situation, allowing a safe return to Earth.

Opposite

Being inserted into the 19-foot-long, 10-foot-in-diameter Gemini capsule was no easy task. Not until Apollo did astronauts have room to move around.

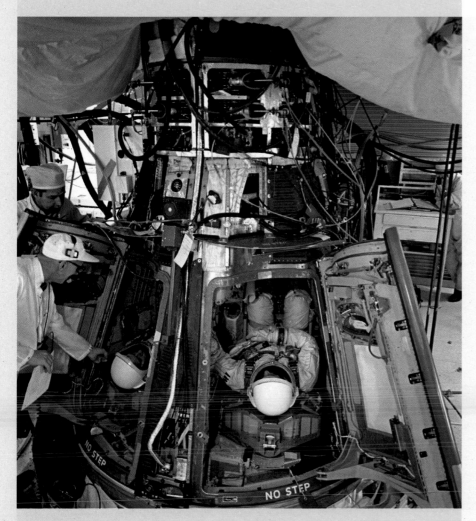

Robert C. Seamans Jr., NASA deputy administrator

It just happened that that was the night of the Goddard [Robert H. Goddard Memorial] banquet. We had [Vice President] Hubert Horatio Humphrey, who was then head of the Space Council, as our speaker. When I left home for the banquet, Neil had docked successfully with Agena, so I was really very pleased, and I got in the car and went over to the banquet. . . . By the time I got there, I was greeted by somebody from NASA saying that we've got a very, very serious problem. And not a lot was then known about it. I got there just before going in to dinner, and I remember feeling that the people there at the dinner ought to know about the situation because otherwise there would be rumors spreading all around. There were enough people in the room who already knew about it.

So I remember standing up and announcing it before well over five hundred people, that we had a serious problem, and people thought I was joking at first. They couldn't believe it. But needless to say, it made the dinner a little more somber than planned and, of course, the press were on this and the news people, and [Walter] Cronkite was on the air, and he was cautiously announcing what was going on and trying to get information as fast as he could. And the information was such that we believed that we would recover [the crew] before Humphrey finished speaking. And I had a handshake—I was sitting right next to [Humphrey]—a handshake that I would tell him just the second we knew. And he started talking, and I was gathering information, and he'd look at me, and then he'd go on for five minutes. You know, even Hubert was having trouble. So finally I gave him the word and he announced that [the crew] was successfully recovered and that ended the evening on a happy note.

Partial air-to-ground transmissions between Gemini 8, Mission Control, and the tracking ship Coastal Sentry Quebec:

Flight [flight director John D. Hodge]: How's it look?

CSQ [James R. Fucci]: We're indicating the spacecraft free. I'm going to call them now.

CSQ: Gemini 8, CSQ CapCom. Comm check. How do you read?

Armstrong: We got serious problems here. We're tumbling end over end. I'm going to disengage from the Agena.

Armstrong: Well, we consider this problem serious. We're toppling end over end, but we are disengaged from the Agena.

CSQ: OK, we get your spacecraft free indication here.

Armstrong: It's a roll or nothing; we can't turn anything off. Continuing . . . in a left roll.

CSQ: Roger.

Flight: CSQ, Flight.

CSQ: Go ahead, Flight.

Flight: Did he say he could not turn the Agena off?

CSQ: No, he said he has separated from the Agena and he's in a roll and he can't stop it. His reg pressure is down to zero, his OAMS [orbital atttitude maneuvering system] regulating pressure.

CSQ: Gemini 8, CSQ.

Armstrong: Stand by.

Flight: Say again.

Armstrong: We are in a violent left roll here at the present time; RCS [rate control system] is armed and we can't fire it. We apparently have a roll of a stuck hand controller.

CSQ: Roger. Did you copy, Flight? They seem to have a stuck thruster. They've initiated the quibs and blown them, but they can't seem to stop it or get them working.

Flight: Did I hear a stuck hand controller?

CSQ: Say again, Flight.

Flight: Did I hear him say he may have a stuck hand controller?

CSQ: That's affirmative, Flight.

CSQ: Flight, this is CSQ. We can't seem to get any valid data here. He seems to be in a pretty violent tumble rate now.

Flight: Roger. What about the Agena?

CSQ: Stand by. We are showing ACS [attitude control system] off on the Agena and we've lost considerable gas pressure.

Armstrong: OK, we are regaining control of the spacecraft slowly in RCS direct.

CSQ: Roger. Copy. Flight, we are getting some summaries out to you right now.

Flight: Roger.

APOLLO TO THE MOON

Earthrise from Apollo 11. The two dozen astronauts who visited the Moon were the first to see our planet from a profoundly new perspective.

President John F. Kennedy's speech to Congress on May 25, 1961, launched Americans to the Moon by giving the Apollo program the stamp of political approval. What most Americans did not know, however, was that the missions had been anticipated by the space community for years. Space stations and manned lunar orbital flights became part of NASA's first ten-year plan in late 1959, during Dwight Eisenhower's presidency and almost two years before Mercury even got off the ground. Although Ike grudgingly approved Mercury as a political necessity, he steadfastly opposed sending Americans on an expensive mission to the Moon, so NASA's planning for it anyway amounted to inspired subversion.

But what was wasting money to the fiscally conservative president went right to the heart of NASA's reason for being: to extend the American frontier to space and send explorers to other worlds. And that ancient barrier was bridged by combining four key elements that launched the most daring voyage in history and left an indelible imprint on civilization: political resolve, outstanding management, extraordinary technology, and Odyssean courage.

Having decided to dramatically disprove Soviet claims that the future of science and technology lay in Marxism, the United States

A Saturn V on its movable launchpad. It took a gigantic rocket to lift three astronauts and all of their equipment to the Moon. The vehicle's engines gobbled fuel at a peak rate of 1,350 gallons a second, and produced as much power as eighty-five Hoover Dams. Perhaps the most amazing statistic of all: The Saturn V launched thirteen times, with no failures.

chose the Moon as a target for its spacefarers and stuck with it through years of painful turmoil at home and overseas. Through it all, NASA administrator James E. Webb and his deputies quickly transformed a loosely organized and far-flung agency into the only organization on Earth that could get men to the Moon and back.

The mission plan and the technology were nothing short of spectacular. There were three possible ways to get men to the Moon: direct ascent, in which one colossal lifter called Nova carried the astronauts on a direct liftoff-to-landing mission; Earth orbit rendezvous, in which two slightly smaller (but huge) Saturn Vs carried the various spacecraft to orbit Earth, where they would be assembled and sent on their way; and lunar orbit rendezvous, in which a single Saturn V would fling the spacecraft modules toward the Moon, where a kind of mother ship would orbit while a descent module landed. Lunar orbit rendezvous was picked because it was less expensive than direct ascent and less complicated than Earth orbit rendezvous.

But it was complicated enough. The three-stage Saturn V, a behemoth developed by Wernher von Braun's team at the Marshall Space Flight Center, had a first stage that alone generated seven and a half million pounds of thrust. The entire lifter towered 364 feet high, including its Apollo spacecraft, and weighed more than five and a half million pounds. The lifter and the two upper stages were designed to send three men on the unprecedented voyage in a three-module spaceship called a stack. A conical command module would hold the astronauts. Behind that was a service module, which had its own maneuvering rocket plus oxygen, propellant, and other consumables. This module was connected to an adapter that held the lunar module, or LM, which in turn was made up of two machines: a descent module that held an ascent module.

The mission was a minutely choreographed ballet. The Saturn V's first two stages would push the stack into one and a half orbits around Earth and be jettisoned. Then the third stage, propelled by a single J-2 engine that developed two hundred thousand pounds of thrust, would fire the stack into translunar "injection" toward the Moon and then fall away. During the three-day voyage, the command and service module (CSM) would separate from the adapter, do a 180-degree turn, move back to the adapter, plug into the lunar module, and pull it out of the adapter. Free of the adapter, all three modules would continue on to the Moon. There, two astronauts would enter the LM, separate from a third astronaut in the CSM, and land on the Moon. Their time on the lunar surface finished, they would climb into the LM's ascent module, blast off and into lunar orbit, link up with the CSM, abandon the ascent module, and head for home, discarding the service module before reentry and splashdown. Voila!

Management of the Apollo program was as impressive as its technology. Every NASA center participated, though the Manned Spacecraft Center, the Marshall Space Flight Center, and the Kennedy Space Center played the major roles. All managers geared their work to a series of rigid reviews, inspections, and finally certifications for every piece of hardware.

Individual rocket components were exhaustively tested, starting in October 1961, when the first stage of the two-stage Saturn I (Saturn V's earliest direct ancestor) was successfully launched with a dummy second stage. In May 1964, a fully functional Saturn I tossed an unmanned test-model Apollo capsule into orbit. It circled Earth fifty-four times and reentered the atmosphere successfully. Additional Saturn I test missions took place through 1965, followed by three unmanned Apollo-Saturn missions that tested launch vehicle structural integrity, launch loads and systems, and stage separations.

Tragedy struck on January 27, 1967, when astronauts Virgil I. "Gus" Grissom, Edward H. White, and Roger B. Chaffee lost their lives in a capsule fire that broke out during a preflight test of AS-204 (later named Apollo 1). A painstaking investigation was conducted and extensive modifications were made on the design of the Apollo capsule.

On November 9, a Saturn V successfully fired an unmanned command and service module named Apollo 4 into orbit. Apollo 5, also unmanned, orbited in January 1968, successfully testing the lunar module. Apollo 6 ended disastrously because of multiple failures, including violent lurching, of the Saturn V.

Then, on October 11–22, 1968, Walter M. Schirra Jr., Donn F. Eisele, and R. Walter Cunningham became the first Apollo astronauts to reach space when they flew Apollo 7 for almost eleven days. They accomplished all of their test objectives and even made a live television broadcast. The stage was now set for Apollo 8 to go to the Moon, which happened on Christmas Eve, 1968. Apollo 9 tested the command and service modules and the lunar module in March 1969.

Apollo 10 reached the Moon on May 21, 1969, and performed lunar module practice exercises that kept the command module and the lunar module separated for eight hours. It returned home on May 26. The dress rehearsal for the most audacious feat of exploration in human history was completed.

Buzz Aldrin photo- graphs his own boot print during the first walk on the Moon. The lunar dust was as fine as talcum powder, but not deep enough to cause problems for landing or walking.

APOLLO 1

D anger always lurked on the way to the Moon. On January 27, 1967, the first U.S. spacecraft tragedy occurred—on Earth. After several hours of dry-run launch practice, Virgil I. "Gus" Grissom, Edward H. White, and Roger B. Chaffee were asphyxiated when a flash fire broke out in their Apollo-Saturn 204 command module.

The sudden fire—virtually an explosion—was caused when an electrical spark ignited the pure oxygen environment in the capsule, which had been pressurized to sixteen pounds per square inch. (Normal atmospheric pressure is 14.7 psi). Oxygen under that kind of pressure is potentially more explosive than gasoline vapor. It was chosen by NASA's Maxime Faget, who headed the manned program's design team, over the objections of North American Aviation, which designed and manufactured the command and service modules. Faget was afraid that ordinary air's balance of 21 percent oxygen and 78 percent nitrogen was potentially perilous. If the sensor that kept the delicate balance failed and too much nitrogen entered the spacecraft, the astronauts would fall asleep and die. Pure oxygen, which had been used in Mercury and Gemini, seemed to be much safer. In addition, it could allow astronauts to better prepare for the pressurization rigors of extravehicular activity.

Ironically, Grissom, who remained haunted by *Liberty Bell 7*'s blown hatch, had insisted that Apollo command module hatches not have explosive bolts. He and his comrades perished in a matter of seconds while rescuers tried frantically to pry open the redesigned hatch.

Following a detailed investigation and congressional inquiry, wire insulation was improved on later capsules and the amount of pure oxygen was reduced. AS-204 was renamed Apollo 1 in honor of the dead astronauts.

▼

Probably the greatest thing a man can say to himself, or have as his philosophy when he has to tackle a tough job, or make a big decision, is the first eight words of the Scout Oath: On my honor, I will do my best. . . .

Roger B. Chaffee, crew member

John D. Hodge, backup flight director

Hodge was in the control center at Houston's Mission Control when the fire broke out. Since it occurred during the countdown test, the control center had no direct responsibility.

We were sitting there and everything was going fine, we had no problems whatsoever. Voice communications were good. Then, bang, there was nothing, absolutely nothing. . . . We had enough data—just a few seconds of data—that indicated that there was an increase in pressure and that was about all. And then there was no voice communication. Very quickly we got news that something had happened. We did not know what, but something had happened. Then, everybody started to play back the data. We had huge rolls of Sanborn graph paper all over the control center floor looking to see if we could find out what happened, expanding it to the maximum amount we could. But there really was not enough data—other than the fact that pressure had gone up—to show what had happened or what had actually occurred to make it happen. We had assumed that it was bad almost immediately.

The Apollo fire took the lives of one of NASA's most experienced astronauts, Gus Grissom (left), the first American to walk in space, Ed White (center), and rising star Roger Chaffee.

APOLLO 1

▼

If we die, we want people to accept it. We are in a risky business, and we hope if anything happens to us it will not delay the program. The conquest of space is worth the risk of life.

**Virgil I. "Gus" Grissom,
crew member**

Jack King, public affairs officer, NASA

It was a very long day. We had all kinds of communications problems between the spacecraft and the control center. I was in the blockhouse, the Saturn 1-B blockhouse at Pad 34. It was a routine test, with all kinds of glitches, and Gus Grissom said at one point, "How can we get to the Moon when we can't even communicate from two miles away?" The test dragged on, and then out of nowhere we heard, "Fire in the spacecraft!"

Everybody was stunned. I started scribbling notes, and listening intently—it was all I could do. Deke Slayton was in the blockhouse as well, and I agreed with Deke that I would hold off on any statement until the three widows were notified. We had to sit there in the blockhouse for about an hour, because sitting on top of the Apollo spacecraft was a 155,000-pound-thrust escape tower. We couldn't send anybody up there immediately because we were afraid the escape tower would explode.

I still have written in pencil the news release announcing that the three astronauts were dead. We got it out as rapidly as we could, considering the circumstances. There was criticism of that at the time, which was understandable on the news media's part, saying I should have announced it immediately, because it was like a presidential assassination. I felt, and I feel secure in it even now, that we did it the proper way.

Eugene F. Kranz, flight director for the Apollo program

I've never seen a facility or a group of people, a group of men, so shaken in their entire lives. . . . The majority of the controllers were kids fresh out of college in their early twenties. Everyone had gone through this agony of listening to this crew over the sixteen seconds. . . . It was very fresh, very real, and there were many of the controllers who just couldn't seem to cope with this disaster that had occurred.

The interior of Apollo Command Module 012 shows the extent of the damage. The fire, most likely caused by an electrical arc, burned hot in the pure oxygen atmosphere.

Burial ceremonies at

Arlington Ceremony,

Virginia, January 1967.

Just twenty-one

months later, Apollo 7

resumed the race to

the Moon.

Lola Morrow, secretary, astronaut office, Kennedy Space Center

While at work on the morning of the test, the crew came in from their quarters for suit-up wearing bathrobes and slippers that I had picked out for them. I had a premonition and I feel that they had a premonition. I said to them, "Are you all right? Is everything OK?" They were just not themselves.

We had a squawk box—an intercom from the test site. I listened to the crew and to everything that was happening. My premonition got worse. I just concentrated on that squawk box. My anxiety grew when I heard Gus saying, "How the hell can we get to the Moon if we can't talk between two buildings?" I tried to stop the test by calling Stuart Roosa, an astronaut at the test site. I was unsuccessful, but he reassured me by saying, "Lola, it's a 'Go.' The problems are ironed out. You can go home."

I called my daughter, Linda, on my way home. "Oh, Mom, you need to call Chuck right away!" she cried out. "There's been a fire! It's on TV, the astronauts!" I called Charles Friedlander, chief of the Astronaut Office. He said, "There's been an accident. How soon can you get here?" I thought, "Oh my God, my premonition came true."

That night I placed calls to the three astronauts' wives for Deke Slayton. I then called the mortician and did whatever else I could do. I left around midnight when the telephones died down. Later, I helped pack up the crew's belongings, which were put on the NASA *Gulfstream* headed for Houston. Days later, along with other co-workers, I watched as the caskets were boarded onto a flight to Arlington and West Point cemeteries.

It was too heartbreaking for me to continue at that office. I volunteered to work with the accident review board. All of my emotions surfaced again when I transcribed the tapes recording the astronauts' final words.

Sam T. Beddingfield, mechanical engineer

It was U.S. Air Force colleague Gus Grissom who convinced Beddingfield to join NASA and explore rocket technology.

What they assigned to me was to take the spacecraft apart, and to take it apart so carefully that we could determine absolutely what made it blow up. I was so busy with that for six months I didn't even have time to think about anything else. The first day of the fire I was supposed to leave my job that night, and on Monday I was supposed to go over to the program office. Well, the fire happened at 6:31 P.M. on Friday, and I finally got home to change clothes about 9 P.M. on Tuesday. That's how busy I was. I lost track of days and nights.

George E. Mueller, associate administrator for manned spaceflight

We spent a great deal of time trying to find the source of the problem, the source of the fire, but literally it was a time bomb just sitting there waiting to go off. . . . Unfortunately, it had to happen then. But we did persevere, and I would say that the good thing that came out of it was that we really understood what causes fires on spacecraft. We redid most of the wiring, not that we knew the wiring was at fault, but rather we redid the wiring on Apollo, and did it much more professionally than the first time around. I think that's probably why the Apollo program was relatively accident-free.

APOLLO 8

Frank Borman, James A. Lovell Jr., and William A. Anders were the first to reach the Moon, but it was for reconnaissance—looking for suitable landing places and other vital details—not a landing. Having left Earth on December 21, 1968, they swung into lunar orbit late on the afternoon of Christmas Eve.

The flight of Apollo 8, which was soon eclipsed by Apollo 11, stands as one of the most important feats of exploration of all time. It not only proved that humans could reach Earth's sole companion, but it gave them—people, not machines—the first close-up look at its ancient, ghostly, battered surface.

With people all over Earth listening or watching live television transmission of their planet, the three astronauts read from the Bible. Borman finished with this historic greeting: "And from the crew of Apollo 8, we close with good night, good luck, a Merry Christmas, and God bless all of you: all of you on the good Earth." The men circled the Moon ten times and then headed home to a tumultuous welcome. Their five-hundred-eighty-thousand-mile odyssey was the longest in history, and having hit nearly twenty-five thousand miles an hour, it was also the fastest.

Lovell said later that the flight gave Americans an uplifting finale to an otherwise dreadful year of assassinations, riots in Chicago, and the deepening quagmire of the war in Vietnam. ". . . [It] couldn't have happened at a better time," he remarked.

▼

Everything worked so well, it was hard to say there were any moments of tension. Of course, we were anxious when we fired the service propulsion engine to slow up and go into lunar orbit. And we were anxious again to fire it going out, but I think those were natural points of concern. Because had they not worked properly, we were dead. But I never really anticipated that we would fail.

Frank Borman, commander

Apollo 8 passes over the Sea of Tranquillity, where the first landing would occur seven months later. The orbit-only mission was added to NASA's plan out of fear that Russia would beat the U.S. to lunar orbit.

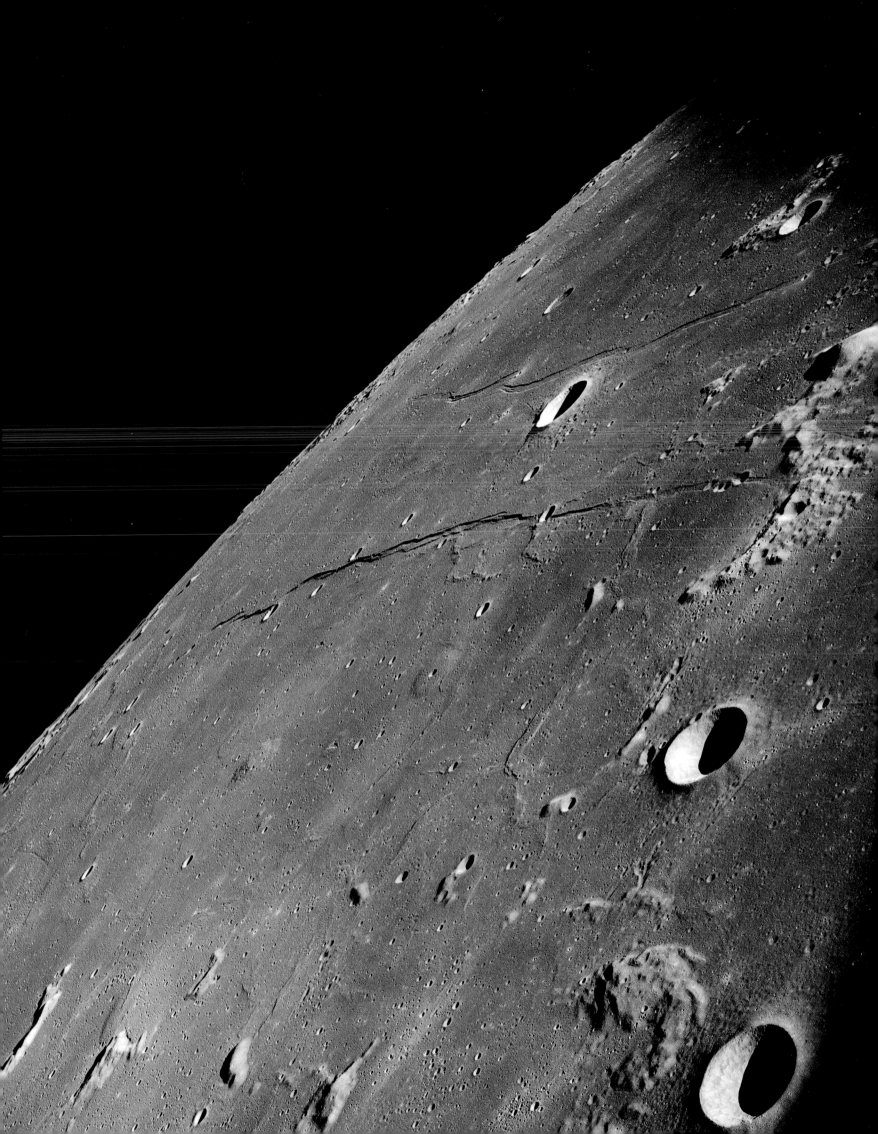

APOLLO 8

William A. Anders, lunar module pilot

We had simulated essentially everything we could think of or anything anybody could think of on that flight, all previous flights, and in centrifuges, in zero-g airplanes, and procedure trainers and that kind of stuff. And yet the very first seconds of the flight were a total surprise to everybody because the Saturn V . . . is a big tall rocket, kind of skinny, more like a whip antenna on your automobile, and we were like a bug on the end of a whip. . . .

For about the first ten, seemed like forty, but probably the first ten seconds we could not communicate with each other. Had there been a need to abort detected on my instruments, I could not have relayed that to Borman. So we were all out of it on our effectively unmanned vehicle for the first ten or twenty seconds. The next most impressive thing was that as we burned out on the first stage . . . we were hitting about six or seven gs. . . . You could hardly lift your arms, you have trouble breathing, but you're not blacked out because the way your blood was flowing from your legs . . . down to your torso.

Then the engines cut off, and just as they cut off, some retro rockets fire to try to move that big first stage away from the second and third stage. . . . So, you go from a plus six g to a minus one-tenth, and the fluid in your ears just goes wild. . . . I felt like I was being catapulted right through that instrument panel. . . . Instinctively, I put my hand up in front of my face, and just about the time I got my hand up . . . the second stage cut in. . . . Whack-o, right onto the face plate with the wrist ring, which left a gash. . . . I thought, "Oh, damn," here I am, the rookie of the flight, and sure enough, here's this big rookie mark. When we got into orbit and I got out of my seat and we took off our suits and each guy handed me [his] helmet to stow and, sure enough, each one of them had a gash in it from the same thing.

But the most impressive aspect of the flight was when we were in lunar orbit. We'd been going backward and upside down, didn't really see the earth or the Sun, and when we rolled around and came around and saw the first earthrise, that certainly was, by far, the most impressive thing—to see this very delicate, colorful orb, which to me looked like a Christmas tree ornament, coming up over this very stark, ugly lunar landscape. . . .

Michael Collins, capcom

I can remember at the time thinking, "Jeez, there's got to be a better way of saying this," but we had our technical jargon, and so I said, "Apollo 8, you're go for TLI" [translunar injection]. If, again . . . you've got a situation where a guy with a radio transmitter in his hand is going to tell the first three human beings they can leave the gravitational field of Earth, what is he going to say? . . . He's going to invoke Christopher Columbus or a primordial reptile coming up out of the swamps onto dry land for the first time, or he's going to go back through the sweep of history and say something very, very meaningful, and instead he says . . . You're go for TLI? Jesus! I mean, there has to be a better way, don't you think, of saying that? [Laughter] Yet that was our technical jargon.

James A. Lovell Jr., command module pilot

On the near side of the Moon along the shore of the Sea of Tranquillity is a small triangular mountain that I named Mt. Marilyn after my wife. Although not officially recognized, the name stuck and was used as a landmark by the Apollo crews. The name is now deeply buried in the vocabulary of the Moon missions.

Left

The Saturn V awaits its lunar voyagers. Among their many firsts, the Apollo 8 crew was the first to ride the Moon rocket, which hadn't yet been tested with people onboard.

Below

The first humans to leave Earth orbit. Bill Anders (left) and Frank Borman (right) never flew in space again, but Jim Lovell commanded the ill-fated Apollo 13.

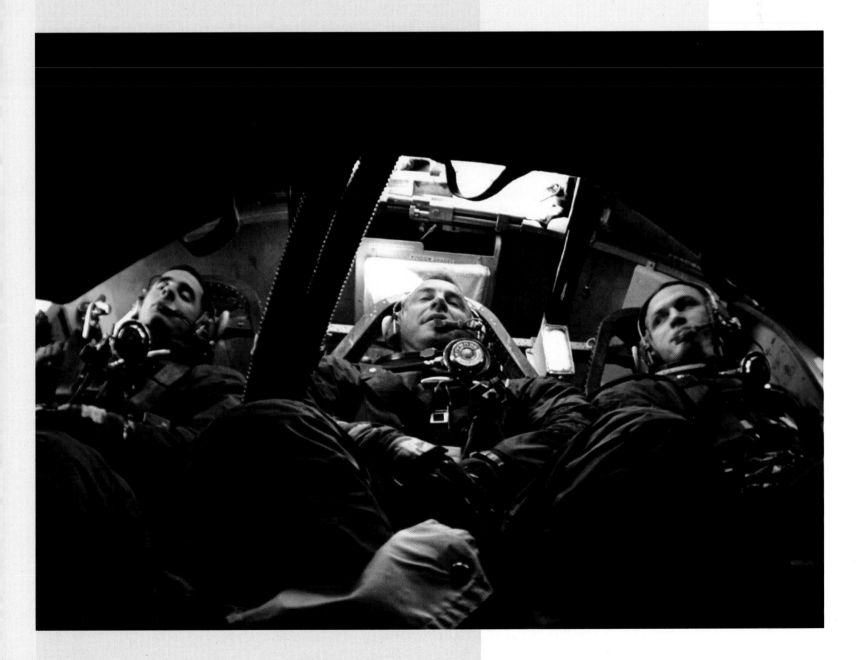

APOLLO 8

▼

The bigger features on the Moon were very easy to recognize, especially the ones on the front side, because we had plenty of photographs and maps. On the back side there were thousands and thousands of craters that no one had ever seen before, except some of the big features photographed by a Russian Luna probe. One was Tsiolkovsky, which is one of the most beautiful craters on the Moon, if you can call a crater beautiful. The other one was the Sea of Moscow.

James A. Lovell Jr., command module pilot

James Holliday, airline pilot

Holliday learned from then–flight director Christopher Kraft the time and coordinates for the Apollo 8 splashdown and offered his passengers the view of a lifetime.

When the pilots bid each month for what kinds of flights they would like, I looked to see what we might have in the splashdown area—and by golly, I found one that would be in the right time and place of reentry. I was a pretty senior captain, and I got the trip on my bid and we got off on time; it worked out beautifully. My first officer and I carefully plotted the Apollo 8 splashdown, and we determined that we would be about sixty-five miles northeast of the area with an excellent view of the capsule coming in. Our aircraft would present no problem to the recovery operation. The last thing we wanted to do was have our airplane hit these guys as they were coming down.

We were cruising at thirty-three thousand feet on an absolutely crystal-clear night. About halfway along the flight we began to see many lights forming a straight line on the surface of the ocean in the splashdown area, so that if the capsule were a little short or a bit long on the splashdown point recovery ships would be close by. We passed overhead and went another sixty to sixty-five miles. By that time it was deadly quiet in the cockpit. We're looking out to the left of the airplane, and suddenly we saw this little red dot underneath the star Capella and we knew that was it. I got on the PA system immediately and told the passengers, "We got it. Look out to the left just over the left wing." All the passengers jumped to an upright position, and then everyone was jammed against the windows.

This dot in the sky was a very dull red. It appeared to be coming straight at us because it was holding its position. It grew a bit larger and a bit brighter, changed from a red to a sort of pink, and then took on an orange color. At that

Right

Up close, the Moon appears battered by impacts and is the color of "dirty beach sand," according to Anders.

Opposite

One of the most famous photographs ever taken. The Apollo 8 crew was surprised at Earth's beauty from space.

point we saw that it was coming in on a line heading toward the splashdown area behind us. The orange color changed to a dark yellow, and a tail became visible— a comet tail. And all of this material remained in the sky; it didn't just die out. So as the capsule proceeded toward the splashdown area, it left this long trail of color that ran from dark red to pink to orange to yellow, with an incandescent light up at the front, which was the capsule. The colors varied so beautifully. I'll tell you, this thing was really moving!

The moment we noticed that it began to move behind us, I put the plane in a very, very shallow bank, so that we could gradually follow it, keeping it right in our windshield view all the way around. Everybody in the cockpit could see it, everybody in the cabin could see it, and we slowly followed this trajectory around until the capsule was just about over the splashdown area. At that point the brilliant, incandescent white light just snuffed out. We expected it to dim somewhat as the capsule slowed down and its temperature decreased, but it didn't. It just went out instantly.

The sad thing is, I had three professional photographers onboard that night, and every one of them had their cameras down in the cargo compartment. I never ever saw anything like that in my life or expect to again, and I didn't have a camera!

▼

To see the earth as it truly is, small and blue and beautiful in that eternal silence where it floats, is to see ourselves as riders on the earth together, brothers on that bright loveliness in the eternal cold— brothers who know now they are truly brothers.

Archibald MacLeish, writer and poet

APOLLO 11

John Kennedy's mandate to land a man on the Moon was carried out at precisely 4:17:40 (EDT) on the afternoon of July 20, 1969, when the lunar module *Eagle* touched upon the Sea of Tranquillity with Neil Armstrong and Buzz Aldrin inside.

While Michael Collins orbited in *Columbia*, the command and service module, Armstrong made his famous proclamation: "Houston, Tranquillity Base here. The *Eagle* has landed."

More than six hours later, Armstrong stepped off the module's ladder and onto the gray, barren lunar surface. The total time that he and Aldrin spent on the Moon was twenty-one and a half hours. Aldrin and his spacesuit weighed three hundred and sixty pounds on Earth, but only sixty on the Moon, allowing him to bound around like a space-borne Pillsbury Doughboy.

The first poke at the lunar surface came when both astronauts tried with great difficulty to plant into the thin dust a collapsible flagpole holding the Stars and Stripes (Aldrin had dreaded the possibility of the American flag keeling over on live television).

During their two-hour-thirty-one-minute EVA, Armstrong and Aldrin collected forty-six pounds of rock and soil samples, took pictures, and left behind scientific instruments, including a miniature seismic station, a laser reflector, and a cosmic-ray detector. On a leg of the descent module, which would also remain, they placed a plaque that read: "Here men from the planet Earth first set foot upon the Moon." It was signed by the three astronauts and President Richard M. Nixon.

▼

I believe that this nation should commit itself to achieving the goal, before this decade is out, of landing a man on the Moon and returning him safely to the earth. No single space project in this period will be more impressive to mankind or more important for the long-range exploration of space.

**President John F. Kennedy,
May 25, 1961**

First on the Moon, July 1969. Neil Armstrong is reflected in Buzz Aldrin's faceplate as he snaps the quintessential Apollo picture. Nearly all the photos on the Moon are of Aldrin.

APOLLO 11

Armstrong, Collins, and Aldrin (left to right) practice their egress (exit) procedures in the Apollo spacecraft a month before their launch. Collins figured the chances of a successful landing and return at about fifty-fifty.

The surf was just beginning to rise out of an azure-blue ocean. I could see the massiveness of the Saturn V rocket below and the magnificent precision of Apollo above. I savored the wait and marked the minutes in my mind as something I would always want to remember.

Michael Collins, command module pilot, prior to boarding capsule

Geneva B. Barnes, secretary, public affairs office, NASA

During the Apollo missions, Barnes assisted in protocol activities at the Kennedy Space Center.

We were at Cocoa Beach a week ahead of the actual launch day and people were already gathering to witness the launch. The evening before launch there were people sleeping on the beaches and in their cars because there were no more hotel rooms. Some spent the night sitting in chairs in hotel lobbies. The motel and restaurant marquees at Cocoa Beach were all saying "Good Luck Apollo 11." There were a lot of well-known people there. You could not help but feel something big was happening and you were just glad to be part of it. My family went down for that launch. I remember when it took off, when it lifted off the pad, I could not help but think, what are they really going to experience once they get there? Are they going to get back? Because in spite of all the things that you heard about the mission being carefully planned and they knew what to expect and everything was going according to the flight plan, I always had a feeling in the back of my mind well, what if . . . I thought that the eyes of the world were focused on Cocoa Beach at the Kennedy Space Center.

Jack King, launch control commentator

During the final three minutes and ten seconds before liftoff I had five different voices [of launch test conductors] in my ear as the different events were clicking down. I knew what I was waiting to hear if everything went right. Just shortly before liftoff, I said, "Ignition sequence start. Five—four—three—two—one. The next cue was "All engines running." Then you had "Commit" and "Liftoff!" Those were the cues you were looking for. I was fine getting through "Ignition sequence start." Then all of a sudden a sixth voice came in my ear, and it was mine. And it said, "My God, we're going to the Moon!" in my mind. And I lost it for a second. If you listen closely, I said, "All engine running." Then I pick up from there.

Well, we celebrated the thirtieth anniversary of the Apollo 11 mission in July 1999, and we brought out about five hundred people who were there on that day. And they played the countdown, and I got up and introduced the launch vehicle test conductor, and all the other people whom I had been hearing "in my ear" during the countdown. And I told the story of how my voice had broken right at the final moment.

Charlie Duke was at my table, and he had been the CapCom in the control center when Armstrong landed. He's the guy who said "Roger, Tranquillity, we're all turning blue here." Well, Charlie actually had said, "Roger, Twangquillity." And he admitted to me that he had had a lot going through his mind, too. Then Neil conceded for the first time that maybe he did not say, "That's one small step for *a* man." In the press conference before this anniversary dinner, he said, "Let's put the 'a' in brackets." So I guess we all owned up, thirty years later.

Below

Mission controllers in Houston explode with emotion at the end of a tense eight-day mission. Succeeding on the first try only sweetened the victory.

A successful lunar landing and Moon walk safely behind him, a relieved Neil Armstrong smiles for the camera while still on the lunar surface.

Charles M. Duke Jr., capcom

For the Apollo program, once the launch vehicle cleared the tower, communications would pass from the launch flight director to the astronaut caspule communicators (CapComs) in Houston.

Just before the landing we were pumped up, and I'd say a little tense. This was something we'd never done before, and it was the first one, and we'd been working on it for more than two years since the Apollo 1 fire. Now we were at the end. I think there was an air of confidence, but also a certain amount of tension in the air as we got closer to the powered descent. . . . During the last minute and a half or so, I was giving the crew a running commentary: "Your fuel's OK, your descent rate's OK," etc., etc. Deke Slayton [Apollo program manager] was sitting next to me, and he punched me and said, "Shut up and let 'em land."

APOLLO 11

▼

We were three individuals who had drawn, in a kind of lottery, a momentous opportunity and a momentous responsibility.

Neil A. Armstrong, commander

Thomas J. Kelly, project engineer, Grumman Corporation

Considered the "father of the lunar module," Kelly was a key designer who headed up the Apollo studies of the LM.

On the Apollo 11 landing there were a couple of hair-raising things. I guess the most dramatic was when Mission Control got a program alarm, which was a pretty obscure condition of the computer software, which could issue dozens of these program alarms. It would just say "program alarm" and then there would be a number, twenty-seven, or something like that. You had to know what the significance of that was. One of these went off about two minutes before touchdown. The crew didn't know what it was and most of the ground controllers did not know what it was. But, fortunately, a couple of the ground controllers had made it their business to study and memorize all of these program alarms that were built into the guidance system software. So thank goodness for that. One of those guys recognized immediately what it was and that it was not a significant problem. So he just said, "Ignore it. Go ahead." That was fairly dramatic.

The other concern was that Armstrong had to take over manually from the guidance computer on final approach because he could see it was driving him right toward a big field full of boulders. So he had to scout around for a better landing spot. And while he was doing that, he was chewing up fuel. So when they touched down, they only had about thirty seconds of fuel remaining, which was about half what they should have had. A normal landing, they should have had about a minute.

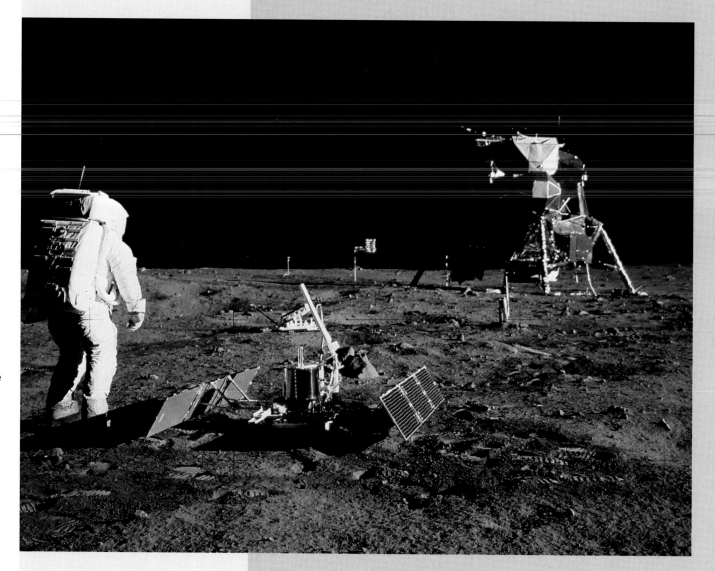

Aldrin makes his way past a seismic experiment, his lunar lander in the background. The astronauts' unprecedented Moon walk experience lasted only two hours, thirty-one minutes—later crews would spend ten times longer roaming the surface.

Aldrin descends the ladder to the lunar surface. This time, there was someone on hand to photograph the "one small step."

Buzz Aldrin, lunar module pilot

The earth didn't look much different from the way it had during my first flight, and yet I kept looking at it. From space it has an almost benign quality. Intellectually one could realize there were wars under way, but emotionally it was impossible to understand such things. The thought reoccurred that wars are generally fought for territory or are disputes over borders; from space the arbitrary borders established on Earth cannot be seen.

Michael Collins, command module pilot

The Moon changes character as the angle of sunlight striking its surface changes. At very low Sun angles close to the terminator at dawn or dusk, it has the harsh, forbidding characteristics that you see in a lot of the photographs. On the other hand when the Sun is more closely overhead, the midday situation, the Moon takes on more of a brown color. It becomes almost a rosy looking place—a fairly friendly place so that from dawn through midday through dusk you run the whole gamut. It starts off very forbidding, becomes friendly, and then becomes forbidding again as the Sun disappears.

Buzz Aldrin

We did find that mobility on the surface was in general a good bit better than perhaps we had anticipated it. There was a slight tendency to be more nearly toward the rear of a neutral stable position. Loss of balance seemed to be quite easy to identify. And as one would lean a slight bit to one side or the other, it was very easy to identify when this loss of balance was approaching. In maneuvering around, one of my tasks fairly early in the EVA [extravehicular activity], I found that a standard loping technique of one foot in front of the other worked out quite as well as we would have expected. One could also jump in more of a kangaroo fashion, two feet at a time. This seemed to work, but without quite the same degree of control of your stability as you moved along. We found that we had to anticipate three to four steps ahead in comparison with the one or two steps ahead when you're walking on Earth.

Neil Armstrong

I couldn't precisely determine touchdown. Buzz called lunar contact but I never saw the lunar contact lights . . . I didn't hear it, nor did I see it. I heard [him] say something about contact, and I was spring-loaded to the stop engine position, but I really don't know whether we had actually touched [down] prior to contact or whether the engine off signal was [on] before contact. In any case, the engine shutdown was not very high above the surface. The touchdown itself was relatively smooth; there was no tendency toward tipping over that I could feel. It just settled down like a helicopter on the ground and landed.

APOLLO 11

The Apollo 11 plaque fixed to the lunar lander shows both hemispheres of Earth and carries four signatures. Rather than claim the Moon for the United States, the astronauts landed "for all mankind."

Yvonne Cagle, astronaut-in-training

On a hot summer night when I was about nine years old, my dad interrupted our play and called us into the house and forced us to watch this news account on TV. It was Neil Armstrong walking on the Moon. I must have spent the next two hours just totally intrigued by the concept of that. I was running in and out between the TV and the full Moon outside, trying to see if I could see this man on the Moon. Did he really exist? After trying to scrutinize this I started to realize I didn't know if I was really seeing the man on the Moon. I might have been, but what was even more fascinating was to imagine what I would have looked like to that man on the Moon. The next thing I thought about was, "Wow. What must it look like from the Moon looking back through space onto Earth?" What would it be like for me? So just in a matter of hours I basically captured the whole vision and dream of what I am living today.

Michael Collins, command module pilot

This is *Eagle* [image at right], or perhaps half an *Eagle* would be better since the landing gear and lower part of the descent stage, of course, remained on the surface. This was a very happy part of the flight for me. I, for the first time, really felt that we were going to carry this thing off at this stage of the game, and it looked like, although we were far from home, we were a lot closer to it than the pure distance might indicate. Neil made the initial maneuvers to get turned around, and then again I did the final docking . . . as Buzz said, the rendezvous was absolutely beautiful. [Aldrin and Armstrong] came up from below—as if they were riding on a rail. There was absolutely no disturbance or off-nominal events during the last part of the rendezvous.

Mike Collins's photo of the lunar lander rising from the surface to dock with his command module.

The Apollo 11 astronauts receive a ticker-tape welcome in New York. The landing was celebrated throughout the world as the fulfillment of a centuries-old dream.

Buzz Aldrin, lunar module pilot

People ask from time to time if we were scared or afraid. Each individual, of course, has to speak for himself. That word, scared or afraid, I guess if you allow that to grab hold of you, it paralyzes your ability to think clearly. . . . En route to the Moon, not much is happening, but you're thinking, What should I be ready to do if things go wrong? . . . Give me clarity of thought. Give me the unfettered mind to be ready to react, to respond.

Bevan M. French, geologist and author, NASA Goddard Space Flight Center

The big triumph of Apollo was that we could use the same techniques we developed for studying the earth to find out about another world. For example, we polished slices of lunar samples till they were thinner than a human hair. Then we examined them under a microscope to determine what minerals were present, how they grew together and how the rock formed. This is a method that has been used to study Earth rocks for more than a century. . . . The first lunar sample I saw with the naked eye looked like a half teaspoon of crushed coal. My initial reaction was that this black stuff was from the wrong place. I couldn't believe it came from something as bright as the Moon. It made a marvelous tinkling sound, like bits of glass, when I turned the vial. That was strange too, because you don't think of the Moon as having a sound.

Julian Scheer, assistant administrator for public affairs office, NASA

I traveled around the world with the Apollo 11 crew after they came back, and the first stop was Mexico. We landed in Mexico City and went from the airport into town. The roads leading into the city were jammed with people. I'd never seen anything like it in my life. We were just stunned. And one of the first handwritten signs we saw was held by someone who had leaned out the window of a village hut. It read, "Apollo 11 Astronauts—This is Your Home." We all saw it at the same time, and it struck us as a great sentiment, said in such a simple way. From day one, the crowds were phenomenal, and we knew we really were onto something. The landing had indeed captured the imagination of the world.

APOLLO 13

I f "The *Eagle* has landed" described the flight of Apollo 11, then "Houston, we've had a problem" expressed the harrowing near-death experience of James A. Lovell Jr., John L. Swigert Jr., and Fred W. Haise Jr. in Apollo 13.

Nearly fifty-six hours into their mission to the Moon, which began on April 11, 1970, they heard a loud bang that NASA later described as an "LO2 tank anomaly." That meant that one of the two supercooled liquid oxygen tanks had overheated, ruptured, and exploded, effectively knocking out the spacecraft's ability to generate power and supply oxygen and water.

Fortunately, the command module *Odyssey*'s reentry system, which had remained intact, had batteries and oxygen to last fifteen hours. With the last of the fuel used to generate power running out, the batteries were employed to keep the inertial guidance system working. Without it, the astronauts would literally have been lost in space. *Aquarius*, the lunar module that Lovell and Haise were to have landed on the Moon, instead became their lifeboat. They utilized its oxygen and power systems to stay alive and its descent rocket to push the command module around the Moon and head it home, where the men landed six days after their mission started.

In a gripping book, *Lost Moon*, Lovell and writer Jeffrey Kluger paid overdue homage to the ground controllers in Houston who worked desperately to save the astronauts' lives. In doing so, they also honored the thousands of men and women who work behind the scenes to make space missions successful.

> All during the ordeal of Apollo 13, we all thought that we were coming back. I guess astronauts have to be optimists. We have to look at all the factors and trust in the equipment that we have and in the people in Mission Control. We knew that we were in deep trouble. We knew that we had to work hard and readjust our procedures, but we had in the back of our minds the idea that we would somehow get back to Earth.

James A. Lovell Jr., commander

Worry time: A group of flight controllers and astronauts, including later Moon walkers Alan Shepard (seated, far right), Ed Mitchell (to his right), and Gene Cernan (standing over Mitchell) wait out Apollo 13's ordeal.

APOLLO 13

Fred W. Haise Jr., lunar module pilot

We'd pulled out a lot of equipment to do a TV show, a kind of show-and-tell thing, and I was putting it all back when the explosion happened. . . . Spacecraft are metal hulls, so there was the sound of a metal bang that reverberated. In the tunnel area—the connection between the two vehicles—there was actually a crinkling of metal sound much like you crinkle a soft drink can in your fingers. So I knew this was real and serious.

I drifted up into the command service module and got into my right couch. There were a lot of lights lit on the caution and warning screen—lights were on for the reaction control system, the fuel cells, the cryo system, the DC bus, AC bus—obviously we'd lost the one tank. It was clearly an explosion of some sort, but of what specific cause, I had no idea. There were failures in multiple systems. That was the confusing factor. There was no single credible failure that we had known that would suddenly cause all these lights to come on at once, except for an instrumentation failure. That's what the ground thought it was for about eighteen minutes before they realized that this was real.

As I went from the lights to look at the meters and things—the instrument panels, the temperature, the pressure and quantity meters for the oxygen tanks—they were all on the bottom for tank two. It meant that we'd lost that oxygen tank, and it became the obvious culprit. But what it also meant was that we were in abort mode. Without looking at the mission rules, which we had for failure modes, I knew it was an abort and we'd lost the chance of landing on the Moon. So my next emotion was this sick feeling over all of my years of training—my backup time on Apollo 8 and 11 and now 13—the training and all the hours and hours down the tubes. . . .

Disappointment got pushed into the background when Mission Control caught up with the fact this was not a caution and warning problem. We got into a troubleshooting routine where I spent most of my time working with the EECOM [Electrical, Environmental, and Communications Systems Engineer] through the flight director and the CapCom to try to isolate things.

Eugene F. Kranz, flight director

Lovell has called down indicating they're venting something and we've come to the conclusion that we had some type of an explosion onboard the spacecraft. Our job now is to start an orderly evacuation from the command module into the lunar module. At the same time, I'm faced with a series of decisions that are all irreversible. At the time the explosion occurred, we're about two hundred thousand miles from Earth, about fifty thousand miles from the surface of the Moon. We're entering the phase of the mission—we use the term "entering the lunar sphere of influence." This is where the Moon's gravity is becoming much stronger than the earth's gravity. And during this period, for a very short time, you have two abort options: one which will take you around the front side of the Moon, and one which will take you all the way around the Moon.

▼

The clincher came when I drifted over to where Jack [Swigert] was sitting in the left-hand seat. Looking past him out the side window, I saw a gas escaping at a high rate of speed, and it wasn't too hard for me to determine, from that indication, and the needle on my second and last oxygen tank, that we were in deep trouble.

James A. Lovell Jr., commander

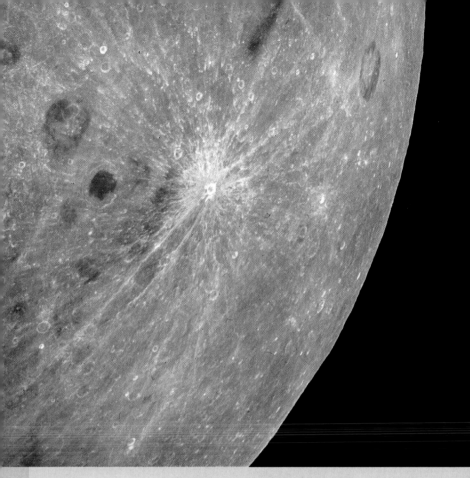

The Apollo 13 astronauts had to satisfy themselves with only views of the Moon, such as this one of a bright-rayed crater on the lunar farside. Still, their trajectory took them farther from Earth than any astronauts before or since.

Eugene F. Kranz

There were emergencies, contingencies, all the way through this process of returning to Earth. There was no such thing as a free ride in this mission. We had to perform a couple of emergency maneuvers because our trajectory was flattening out. We didn't know why. We had to correct that. The crew was suffocating. We had to invent techniques of using the square chemical scrubbers we used for the air from the standpoint of the command module and be able to adapt those over to the lunar module.

Thomas J. Kelly, lunar module project engineer, Grumman Corporation

The lunar module was reconfigured to supply the necessary power and consumables for the crew's survival.

We needed to know very precisely how fast the different consumables were being used, and by what. That applied to everything: electrical power, water, oxygen. . . . What we were doing was digging out the test records on the specific pieces of equipment that were up there on the LM. We wanted to know exactly how much current a particular unit drew the last time it was tested in the lab. We wanted to get it down to a "gnat's eyelash." We found out very accurately what each piece of equipment consumed, and we were doing all of the calculations to see how long we could last. And it wasn't a pretty picture. It turned out we were going to be very short, particularly on power and water. And we were going to have to shut down an awful lot of equipment that we did not want to, but we did not have much choice.

Fred W. Haise Jr.

The lunar module power-up was one of the pressure points I felt during the flight. We had to do it quickly. We didn't have an activation procedure that was a quick activation. The book we had was one that was prepared for landing, so we had to kind of ad-lib our way with Mission Control, deleting certain steps and that kind of thing. It was interesting when Jim and I got back on the ground and went into a simulator in a nice, calm environment on Earth. We could not match the time of the power-up that we did in flight. I think the adrenaline was flowing a bit more then.

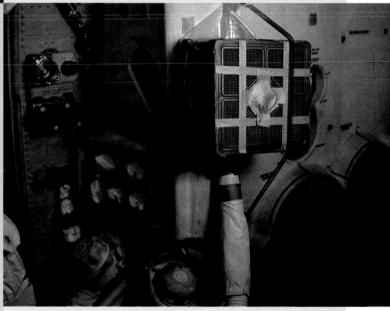

The jury-rigged air "scrubber" that saved the three astronauts' lives. Designed by engineers in Mission Control, it kept the lunar module's air breathable.

Opposite Prior to reentering Earth's atmosphere, the returning astronauts get a good look at their damaged service module. The explosion had ripped away an entire panel.

APOLLO 13

Charles M. Duke Jr., backup crew member

In the first twenty hours after the accident I wasn't sure they were going to make it. But that certainly wasn't communicated to the crew. It was all positive: "We're working on it. Hang in there guys, we'll have this solution for you." And sure enough, we got smarter and smarter, and the lunar module performed better and better. So when they started back around eighty hours into the flight, and were on their way home, my thoughts changed to, "Unless we really screw it up down here or up there, we're gonna make it."

Eugene F. Kranz, flight director

As we were approaching the final phase of entry, the procedures weren't coming together quite as nicely as we would have liked. The crew wanted to see how we intended to accomplish this final sequence. The basic problem we had was, we had a command module that was our reentry vessel; it had the heat-shield, but it had only about two and a half hours of electrical power lifetime. We had the service module, which is where the explosion had occurred; it was virtually useless. We had the lunar module, [which] was attached on the other end of this stack through a small tunnel, and that was our lifeboat. We had to come up with a game plan to move this entire stack into an attitude where we could separate all three pieces in different trajectories so they wouldn't collide with each other in entry. Then the crew had to evacuate from the lunar module lifeboat at the very last moment, power up the command module, get its computer initialized, separate the pieces, and get into attitude for entry. So, this is the game plan we were coming up with. And we didn't really get all the pieces put together and get them verified in simulators until about ten hours prior to the time that we had to execute this plan.

> The command and service module was not designed to ever be turned off and then back on in flight. There was no activation procedure for that. It had to be invented by Mission Control. It had to be their biggest task. When we used other procedures before flight and tried them out in the simulation, we'd go through three or four revs to get it right. Even in sims we would find errors. . . . The power-up was kind of miraculous. And it was perfect. There was not one error.

Fred W. Haise Jr., lunar module pilot

The drama of Apollo 13 gripped the entire world for four days before ending with a splashdown in the Pacific ocean on April 17, 1970.

Eugene F. Kranz

During Earth reentry a blackout period occurs, in which ground-to-spacecraft communication is lost.

For each controller during blackout, this is an intensely lonely period. . . . The crew's on their own . . . left with the data that you gave them—maneuver data, attitude information, all of these kind of things. And each controller's going back through everything they did during the mission [asking themselves], "Was I right?" That's the only question on their minds. . . . And every eye is on the clock on the wall, counting down to zero. When it hits zero, I tell [Joseph] Kerwin, "OK, Joe, give them a call." We didn't hear from the crew after the first call. We called again, and we called again. We're now a minute since we should've heard from the crew, and for the first time in this mission, there is the first little bit of doubt that's coming into this room that something happened and the crew didn't make it. But in our business, hope's eternal, and trust in the spacecraft and each other is eternal. So, we keep going. And every time we call the crew, it's "Will you please answer us?" And we were one minute and twenty-seven seconds since we should've heard from the crew before we finally get a call—a downrange aircraft has heard from the crew as they arrive for acquisition of signal. Almost instantaneously from the aircraft carrier, we get: "A sonic boom, Iwo Jima. Radar contact, Iwo Jima." Then we have the ten-by-ten television view. And you see the spacecraft under these three red-and-white parachutes, and the intensity of this emotional release is so great that I think every controller is silently crying.

In Mission Control, the unfortunate thing is—I guess it's necessary—you can never express an emotion until well after a mission is over. . . . These guys are in the warm air of the South Pacific. They're home. They're alive. And in Mission Control, our job isn't done until we've handed over the responsibility to the carrier task-force commander. It is only when that is accomplished that we can start this internal celebration.

Haise, Swigert, and Lovell onboard the rescue carrier. Relief at making it home overshadowed their disappointment over missing the Moon landing.

▼

Apollo 13 does offer the best example of all the ingredients that make for success: the right skill base, the right training, the right teamwork and trust within the team, and the right leadership.

Fred W. Haise Jr.

APOLLO 17

The flight of Apollo 17 in December 1972 lowered the curtain on the Moon program with a flourish.

With Ronald E. Evans circling in the command module America, astronauts Eugene A. Cernan and Harrison H. "Jack" Schmitt (a geologist and the only scientist to visit the Moon) spent a record seventy-five hours in the Taurus Mountains on the most science-oriented Apollo mission. They collected 243 pounds of rock and soil samples, set up the last in a series of automated research stations, conducted more experiments, and traveled farther than any of the ten astronauts who preceded them.

To all of that can be added the distinction of having the first auto accident on another world. Cernan and Schmitt drove an electric Lunar Roving Vehicle (or simply LRV) twenty-two miles at a top speed of nine miles an hour during one of their three excursions. Cernan not only managed to snag a hammer on the rover's fender, ripping it open, but he dented its tires by driving over rocks.

Cernan was also the last man on the Moon. Before he climbed aboard the ascent module on December 14, he left a plaque on the descent stage that read: "Here Man completed his first exploration of the Moon in December 1972 A.D. May the spirit of peace in which we came be reflected in the lives of all mankind."

America, the Apollo 17 command module, is on display at the Johnson Space Center in Houston. The damaged LRV remains where Gene Cernan left it.

> The dramatic structure as explained to the American people was: Man. Moon. Decade. That's a thriller. Speaking to you in Hollywood terms, it's got everything that you could possibly want for a movie. But what every good screenwriter knows is, once you arrive at the end of the story, the movie is over.
>
> **Michael Gray, author**

The last Moon rocket waits on the pad, its target looming in the background. Three and a half short years after the first lunar landing, the Apollo era was nearing its end.

APOLLO 17

Eugene A. Cernan, commander

A lot of people get fooled by the words "Mission Control" and think that the ground controlled our spacecraft. Really, it was our "management advisory office" back here on Earth. With telemetry they could read all kinds of information that we didn't have in the cockpit, largely because of weight constraints and the complexity of the spacecraft; but we were the ones who were flying the spacecraft. We had computers to do a lot of things for us, and we got a lot of information from the ground; but, ultimately, it was the crew that had to tell the spacecraft what to do. In the U.S., the battle for crew control of the spacecraft had been won in the early days of Mercury. Now, I could have put the spacecraft in a configuration where the computer could have done the landing, but it wouldn't have known if it was going to land on a rock or in a hole. And there's no pilot in the world who would have traveled a quarter-million miles only to let a computer do the landing.

Thomas J. Kelly, lunar module project engineer, Grumman Corporation

On the last LEM [lunar excursion module] to be launched, the Grumman people down at the Kennedy Space Center were literally working themselves out of a job because after the launch, they were going to close up the operation down there. Some of them were going to get jobs back in Long Island [New York], but most of them were just being let go. Now, NASA was pretty concerned about that. They thought, "Gee, what kind of morale are these guys going to have with that situation?" Our leader down there, George Scurla, told Colonel Patrone—Rocco Patrone, the NASA leader—"Don't you worry about my Grummies," he said. "They are going to do just fine." One of the things that they did was to make a poster that they sneaked up to the LEM level on the launch stack. This was against the rules but they did it anyway. It was a big poster that they all had signed. And they put it where the astronauts could see it before they got into the command module, when they were going on the mission. It said, "This may be our last LEM but it will be our best." And it was the best, with no flight anomalies.

▼

Change on the Moon is even slower than we once thought it was. The astronauts' footprints are still there and they will probably still be there for several million years to come.

Bevan M. French, geologist and author, NASA Goddard Space Flight Center

Harrison "Jack" Schmitt rounds a large boulder named "Tracy's Rock," after the young daughter of his crew- mate, Gene Cernan. Schmitt was NASA's first scientist-pilot.

Cernan driving the lunar rover in the valley of Taurus-Littrow. The specially built Moon buggy added mobility to the last three Apollo expeditions.

APOLLO 17 **101**

Eugene A. Cernan

The rover just didn't move very fast. It wasn't made to go very fast, but it could sure save a lot of energy. You could pack a lot of gear and save a lot of walking time, and it allowed us to cover territory and ground we never would have covered otherwise. I think probably we averaged eight to twelve kilometers [five to seven and a half miles per hour], and I think we hit a speed record of fourteen or fifteen kilometers coming downhill. . . .

Of course, it could still be very sporty driving. When you start getting one wheel in a crater, or you hit a bump, or you start driving on the side of a hill, you really do appreciate the one-sixth gravity. If you hit a bump, you'd go u-u-u-u-p and come down. It wouldn't just be bing-bang, up and down. And on the side of a hill, you would start thinking that the rover was just going to tip over and fall over on you. You didn't feel like there's much holding you to the surface at all. . . . I remember many times when we were driving on the side of a hill and Jack [Schmitt] commenting that we were going to roll over. And indeed, because you have less gravity holding you down on the Moon, that was not a false perception; it was fact. You had to be careful on the side of a hill, because if you hit a bump with an uphill wheel, you could lift the thing off the ground and possibly become unstable and tip over. Of course, the situation was more obvious if you were in the downhill seat and I tried to keep Jack on the downside. It was much more comfortable on the uphill side; and that's a commander's prerogative when he's driving the rover.

Joseph P. Allen, capcom

As the mission's Goodnight CapCom, Allen concluded the daily communications with the crew.

I was a person who had come out of the Midwest. I'd never made a phone call across an ocean before. That's the way things were then. One wrote letters. This was in the late 1960s, there was no e-mail and, although transatlantic phone calls were not unknown, they were horribly expensive. Shortly before then, I'd been a student with very little money and just didn't make those kinds of calls. And suddenly, I had an occasion where I was talking to men on the Moon for hours on end. It was really quite bizarre.

APOLLO 17

Cernan salutes the flag. In the background are the lunar lander and the parked rover. Sticking out of the leg pocket of the commander's spacesuit is a geology hammer.

Harrison H. "Jack" Schmitt, lunar module pilot

We were really focused on the mission and didn't have a great deal of luxury to worry that it was the last one. . . . The first time I began to think about that in any serious way, other than regret, was after we had rendezvoused with Ron [Evans] and were closing out the lunar module. We were interrupted, if I remember the sequence correctly, at that point with Jim [James] Fletcher [NASA administrator] reading President Nixon's statement to us, which included words to the effect that we would be the last to visit the Moon this century. I remember that upsetting me greatly, one, because I thought it was an unnecessary remark, even if it were true, and two, I hoped that we would prove him wrong.

Joseph P. Allen, capcom

By Apollo 17 there was a certain nostalgia that began building up. But at the same time, all of the operational people—those who fly the mission, those who control the mission—are very focused during mission time. There's not a whole lot of free time to reflect. You sort of headed into it thinking, "Oh boy, this is the last one, let's not mess it up." So, maybe there's a bit of an added pressure. . . . I think that at the point the crew returned safely, there was a considerable sigh all around. And then people began to do a lot of thinking, "OK, what do we, or what do I as an individual, do next?" A number of very professional flight controllers and operational people began to think about leaving what had been, in hindsight, quite an extraordinary effort.

Eugene A. Cernan, commander

As the last person to have walked on the Moon, Cernan describes his final steps up the ladder to the Challenger *lunar module.*

I remember that moment vividly. We'd been on the Moon for three days and we certainly didn't want to go. I looked over my shoulder and saw my last footprint, knowing that it was a moment no one could take away. It would be the last time anyone would see this valley [Taurus-Littrow] on the Moon for a lot of years into the future. To top it off the earth was sitting just above the mountains in the southwestern sky. It was very special. I was excited.

Bevan M. French, geologist and author, NASA Goddard Space Flight Center
One of the sampling techniques used to obtain detail about the stratigraphy of the lunar landing site included taking a drill core sample of the soil.

The deep-core sample proved the changelessness of the Moon. The Moon had a very violent early history with eruptions of lava and impacts of giant meteorites. Then, four billion years ago, the solar system ran out of big objects and the volcanoes quieted down. So the rubble layer formed a thousand times slower than the slowest sediment collection on Earth, which is in the deepest ocean basins. If you laid the core sample on a big dining-room table and walked just a few steps along it you would be moving from the present day to more than a billion years ago with virtually no changes in the sample. If you went back a billion years on Earth, your sample would look very different.

Harrison H. "Jack" Schmitt

Apollo 11 began the process of understanding the evolution of the Moon and, indeed, the early evolution of the Earth. From the Moon, we gained information about the early history of the Earth that would have been impossible to gain from the Earth itself. Through this remarkable effort on the part of national leaders, managers, engineers, workers, industry, a worldwide community of scientists, and the American people and their families, and through the other Apollo and lunar missions, we now have looked with new insight at our own planet and other terrestrial planets. . . . The Apollo explorations were of incalculable value in adding the reality of known materials and processes to the interpretation of data from subsequent automated explorations of the Solar System—something we also, I think, tend to forget. . . . what would we have known about Mercury and Mars and Venus and other terrestrial-type planets without the detailed understanding we achieved about the Moon?

Below left
Following its night launch, Apollo 17 went into translunar injection with a fully lit Earth underneath it, allowing astronaut Schmitt to capture one of the world's most widely seen images.

Cernan (left) and
Schmitt closed a final Moon walk with the words, "We leave as we came, and, God willing, as we shall return, with peace and hope for all mankind."

▼

You look back at the earth, which is surrounded by the blackest black you can imagine, and you get a sense of the endlessness of time and space. It's overwhelming, unrealistic. You grasp for your identity at that moment. You just want to have everyone standing next to you, feeling what you feel.

Eugene A. Cernan

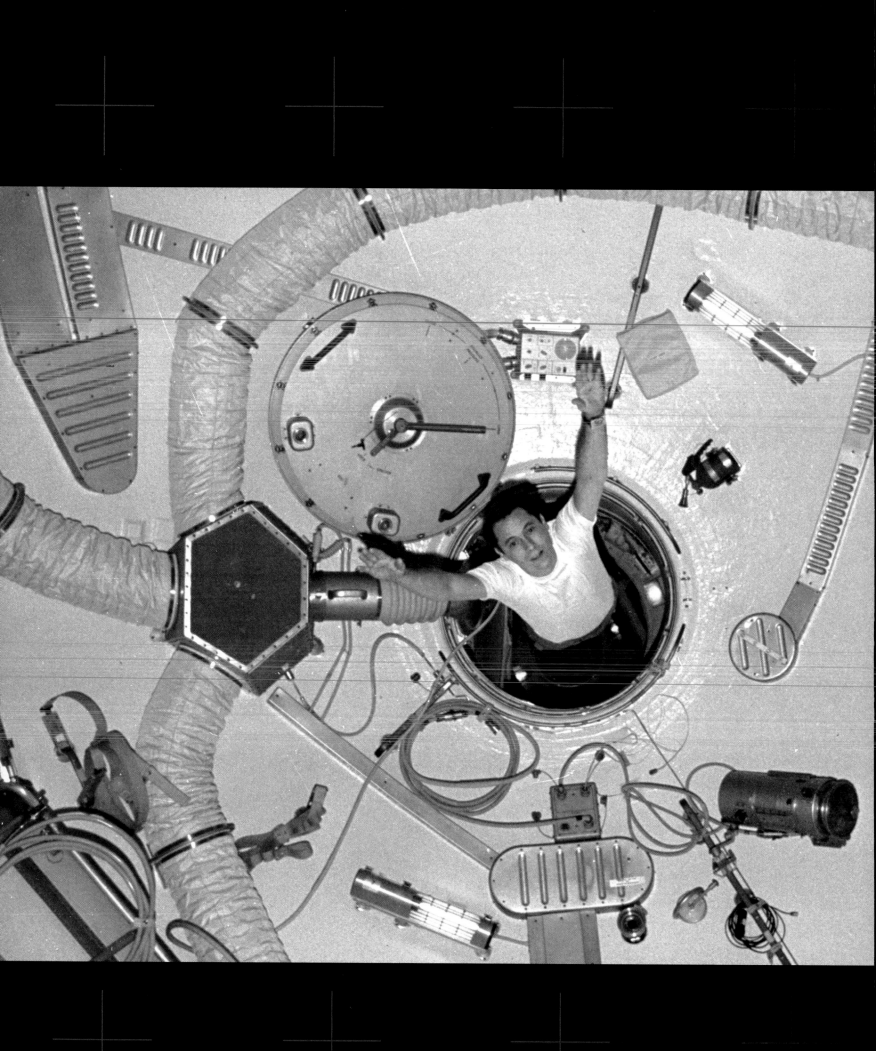

THE POST-APOLLO ERA

POST APO PO

The landing of Americans on the Moon in the most daring and dramatic feat of exploration in history created a monumental paradox for NASA: It was an achievement so supremely impressive that it was impossible to equal. In addition, a fickle public's attention span started declining after Armstrong, Aldrin, and Collins had shown that humans could land on the Moon and return safely. Where ordinary Americans were concerned, the point didn't have to be made again, and again, and again . . . . So despite what Cernan and Schmitt accomplished scientifically on Apollo 17, the mission was pretty much of a sleeper for the public.

Three less-dramatic human spaceflight programs were in the works, even before Cernan closed the door of his ascent module on December 14, 1972, and headed home. One was the start of design work on the space shuttle, which had been approved by President Nixon in January 1971. The other two, called Skylab and the Apollo-Soyuz Test Project, depended on the then-venerable Saturn boosters and "Moon surplus" Apollo command and service modules. The first was scientific; the second was unabashedly political.

Ed Gibson floats through the airlock module hatch during his Skylab 4 mission in 1973. Gibson and his crewmates Gerald Carr and William Pogue spent eighty-four days in orbit—still the longest U.S. space mission.

Skylab

Skylab's genesis was a study NASA did in 1959 for a permanent manned orbiting laboratory. The study was shelved when the race to the Moon got under way. But in 1963, more than six years before the first manned Moon landing, the Manned Spacecraft Center in Houston revived the station idea as a way to use surplus Apollo hardware. Two years later, Wernher von Braun's Marshall Space Flight Center began considering ways to use the Saturn V's third stage, called Saturn IVB, and whole Saturn Vs. Marshall favored a "wet" workshop, meaning that a fueled Saturn IVB would be orbited and then converted into a laboratory by an Apollo crew working in space. Ultimately, however, a less-risky "dry" workshop plan was adopted, in which a Saturn IVB, fully modified on the ground instead, would be lifted to orbit by a Saturn V. And it would carry an Apollo Telescope Mount for solar observation.

Skylab's primary purpose was to study the long-term effects of weightlessness, radiation, and other phenomena on astronauts in preparation for the colonization of the Moon and an expedition to Mars. The station was launched on the last operational Saturn V Moon launcher on May 14, 1973, and was visited by three teams of three astronauts who reached it and returned to Earth in Apollo command modules.

Skylab's heart, the Saturn IVB, was converted into a forty-eight-foot-long workshop with living quarters and a laboratory. Relative to Apollo capsules, the workshop was the Ritz Hotel, with a refrigerator, stove, private toilet, and the first shower in space (it leaked). Visiting astronauts worked in their shirtsleeves and dined on prime rib, potato salad, scrambled eggs, and ice cream, not gunk in tubes.

More than a hundred experiments were performed on the successively longer missions, including medical, Earth resources, solar physics, and materials processing. A number of experiments, some involving plant growth and immunology, came from students. Meanwhile, the Soviets were accomplishing the same thing with their own Salyut stations. And whatever the rhetoric of the Cold War, scientists on both sides were exchanging their newfound knowledge.

The Apollo-Soyuz Test Project

While en route to meeting Apollo 11 astronauts Aldrin, Armstrong, and Collins after their splashdown, NASA administrator Thomas O. Paine, who succeeded James E. Webb, told President Richard Nixon that the Moon landing had established U.S. space supremacy once and for all and that it was time to "stop waving the Russian flag" as justification for the space program and instead justify it for more fundamental reasons. He even suggested cooperation with the Soviet Union; in effect, the victor would magnanimously extend a hand to the loser of the Moon race. It was a very prudent and farsighted strategy. Competition was a negative, not a positive, reason to explore space. More to the point, no competition lasted forever, and Paine did not want the U.S. civilian space program grounded when the Cold War abated.

Apollo-Soyuz marked the true beginning of working together, not so much because of what happened in space, but because of what it took to get there. American and Soviet engineers spent five years working closely in six groups that operated in Houston and Moscow, coordinating rendezvous methods and compatibility, guidance and control, design of a docking module,

While en route to meeting Apollo 11 astronauts Aldrin, Armstrong, and Collins after their splashdown, NASA administrator Thomas O. Paine, who succeeded James E. Webb, told President Richard Nixon that the Moon landing had established U.S. space supremacy once and for all and that it was time to "stop waving the Russian flag" as justification for the space program and instead justify it for more fundamental reasons.

Seamstresses from a sail-making company in New Jersey were enlisted to help make an emergency sun parasol for the Skylab space station after it lost its thermal protection during launch.

astronaut-cosmonaut training, and more.

Out of this space-age anvil chorus of engineers came a joint agreement on mission operations; life-support systems; safety, flight control, tracking, and communication techniques; and interactive crew procedures. The spacefarers more or less learned each others' languages and became familiar with their respective flight procedures and spacecraft. And the spacecraft themselves were connected by a unique international docking module—a transfer tunnel—that was carried to orbit with the Apollo command and service modules.

In the process, years of mutual suspicion, differing design philosophies, and language barriers were overcome. The Russians were so worried about U.S. espionage that they flew Americans into the Tyuratam cosmodrome only at night to restrict their view of the immense facility (which was regularly photographed by spy satellites). Meanwhile, some congressmen grumbled that Apollo-Soyuz amounted to giving away U.S. space technology to Communists.

Later, when relations between the two superpowers soured, there were further complaints in Congress that Apollo-Soyuz had been merely a political stunt. But the spirit of cooperation would outlive the Cold War.

PROJECT SKYLAB

Skylab 1, America's premier space station, launched on May 14, 1973. Sixty-three seconds after liftoff, the one-hundred-ton flying workshop ran into serious trouble when vibration tore loose its meteoroid shield, which was also designed to protect Skylab from the Sun's heat. The shield, in turn, caused the partial deployment of one of the solar "wings," which was then blown into space by the exhaust plume of the second-stage retro-rockets. A strap of shield debris jammed the second wing. As a result, virtually no power to the station could be generated and Skylab's internal temperature rose too high for human survival.

During the next ten days, scientists, engineers, astronauts, and management personnel at the Marshall Space Flight Center and other NASA centers worked around the clock to make the station habitable.

On May 25, 1973, three astronauts on a mission called Skylab 2 (Skylab 1 was the launch of the station itself) arrived in an Apollo command and service module and rigged a parasol to protect the station's exposed areas from direct sunlight. They lived and worked on the two-level station for twenty-eight days and were followed by the crew of Skylab 3, who stayed for fifty-nine days, and finally by Skylab 4 astronauts, who set a record of 1,214 orbits in eighty-four days.

By the time of the last crew splashdown on February 8, 1974, more than one hundred experiments had been performed during an unprecedented 171 working days in orbit. Skylab continued to orbit the earth unoccupied until the pull of gravity eventually caused it to make a fiery plunge into the night sky on July 11, 1979.

▼

With a glance out the window you can see only the wind-swept deserts in North Africa and know where you are, see cloud streaks over South America and know you're over Brazil, or see ocean currents and know where you are over the Atlantic. You get to know the earth like the face of an old friend.

Edward G. Gibson, scientist-pilot, Skylab 4

The Skylab 3 crew, inside an Apollo-type command module, inspects its new home before docking with the station in July 1973. The windmill-like solar arrays are for generating power.

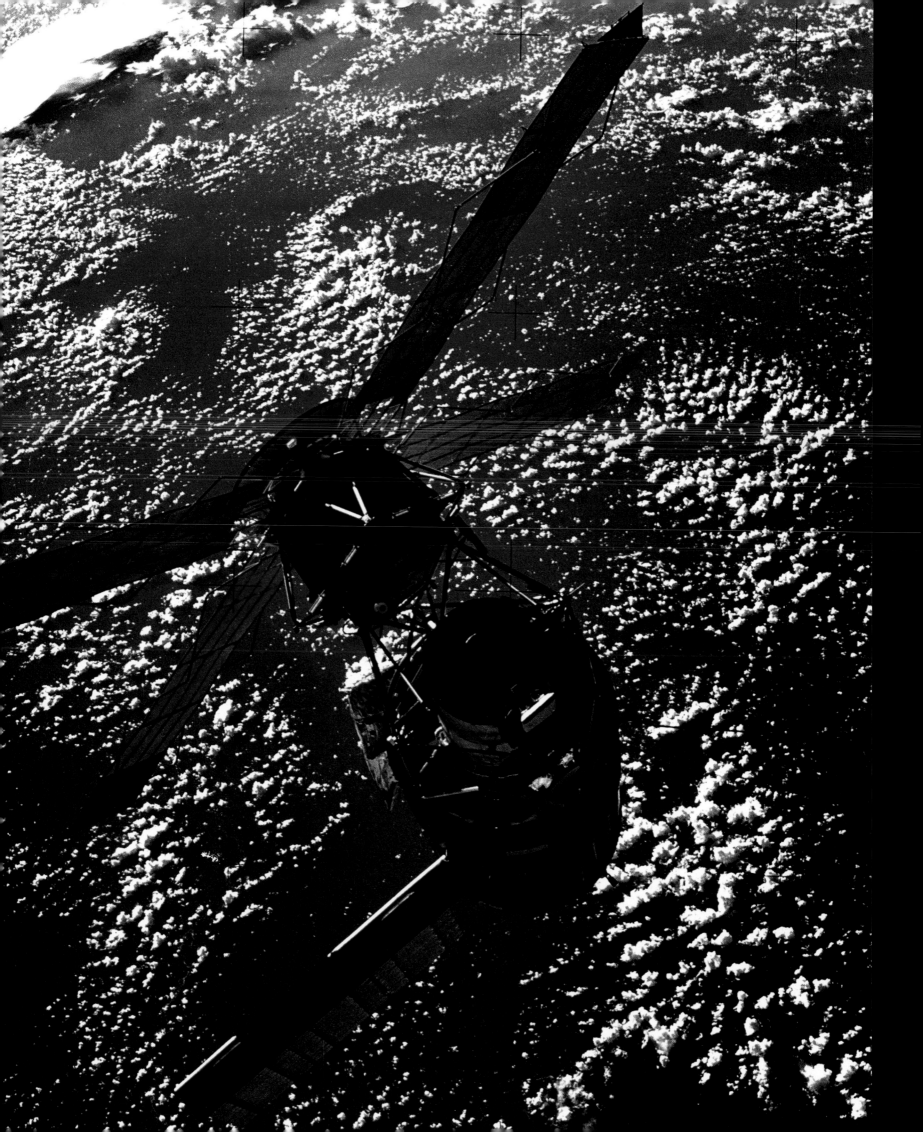

SKYLAB

Owen K. Garriott, scientist-pilot, Skylab 3

There was a lot of preparational work, more than we ever imagined would be required. For example, none of us were really solar physicists, and yet solar physics was one of the prime experiment areas, in which we would spend at least a third of our time looking at the Sun, making recordings and working jointly with the principal investigators on the ground. So we had classroom lectures. In fact, Ed Gibson and I had to set up the whole lecture program. It was almost like creating an academic course in solar physics, which all of the Skylab astronauts went through.

Edward G. Gibson, scientist-pilot, Skylab 4

The equipment to save the station was quickly designed and tested, including the Mylar sun parasol intended to lower Skylab's internal temperature and the cutting tools required to free the remaining jammed solar panel.

We could see from the satellite data what the problems were on Skylab. We got a number of people together and tried to figure out what we ought to be doing on EVAs to make those fixes. We then went over to Marshall Space Flight Center, worked in their water tank [Neutral Buoyancy Simulator] developing procedures and hardware.

We would come out of the water tank, decide what we needed, and then have ourselves, the center director, and the program manager all standing around the lathe while the guy was working on the flight equipment. We'd tell him what to do, evaluate the work, and make changes right on the spot. NASA has never seen anything like it before or after, where we just got things done in the shortest time possible. Now, it would probably take NASA months and months to do it. We were doing it in a matter of minutes.

▼

Mother Nature can't even grow a perfect crystal, but we think we can do it outside the influence of gravity.

Charles "Pete" Conrad, commander, Skylab 2

Astronauts Russell L. Schweickart and Edward Gibson experiment underwater with cutting tools and methods to free a solar wing on a mock-up of Skylab.

Opposite

Skylab 4 scientist-pilot Ed Gibson (left) and mission commander Gerald Carr are surrounded by trash bags inside the Skylab Orbital Workshop. Compared to earlier spacecraft, the station was cavernous.

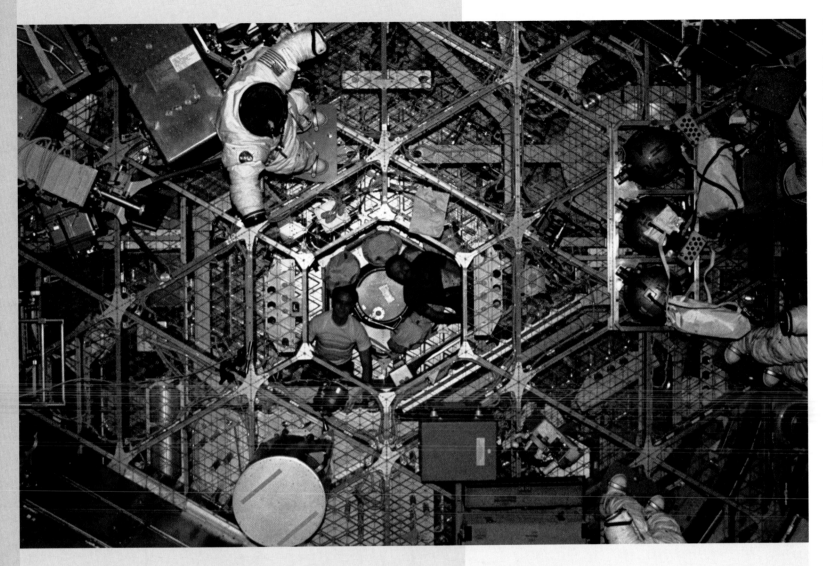

Owen K. Garriott

The days between launch and May 25, when the first crew went up were the finest days of NASA's existence, with the possible exception of the three or four days of the Apollo 13 return. The Apollo 13 return was principally a Johnson Space Center endeavor. Skylab was an agency-wide effort. Houston, Marshall, Kennedy, and several of the other centers, all assisted in figuring out the problems and devising schemes to fix them, which enabled the mission to go ahead.

Edward G. Gibson

Docking is relatively straightforward as long as you've got a good knowledge of where you are, where the spacecraft is, and how fast it's going. You've got a computer onboard that can calculate it. It turns out that we also did backup calculations based on just the distance from the spacecraft and how fast we were coming toward it, called range and range rate. From those alone I used a little HP [Hewlett Packard] calculator, along with some charts, to calculate the maneuvers we had to make in case our computer went down.

When we got up there and saw Skylab, we were just all eyes pressed against the window. We finally came in to do the docking, and I think Jerry [Gerald P.] Carr was trying to be a little too gentle with it. He drove it in and bounced off. There were beads of sweat on Jerry's forehead, and I pulled his chain a bit about being a tough Marine. On the next attempt, he made sure we made it.

Owen K. Garriott

During Skylab 3, Garriott and his crew members Commander Alan L. Bean and Pilot Jack R. Lousma broke the long-duration stay record with fifty-nine days lived in space.

We essentially had minimal friction, if any, among ourselves. I think that with crews as small as three, when they train and work together for as long as Apollo and Skylab crews had done, these problems are almost nonexistent. However, once you get to a crew of six and more, with both genders and two or three nationalities represented, and you want to fly for three months at a whirl, if there are no sociological problems you are extremely fortunate. So I think that for the International Space Station, NASA is going to have to consider that very carefully.

SKYLAB

The first U.S. space station, Skylab was built partly from leftovers. Its giant cylindrical Orbital Workshop was a hollowed-out Saturn rocket stage.

Edward G. Gibson, scientist-pilot, Skylab 4

Even though we did have some problems with the design, by having flight crews onboard to fix those things, we could still achieve mission success. If Skylab was unmanned, it would have been lost. Thus, it showed one real value of having humans in space. Of course, if we didn't have the mechanical problems, we wouldn't have had to prove that, but that's not the real world. Not everything that's built works as intended, as you know from cars and computers. You need a human around to make those fixes when required, and luckily we were able to make all the fixes required.

Alan L. Bean, commander, Skylab 3

Being in Skylab for two months was much more difficult than going to the Moon. It was a greater personal challenge, and at the end I felt like I had given more of what I had to give as an astronaut. It took more self-discipline and skill to be commander of Skylab for two months than it did to be lunar module pilot and go to the Moon.

Edward G. Gibson

The Apollo Telescope Mount (ATM) was the largest piece of scientific equipment on Skylab. Free from the interference of the earth's atmosphere, it returned invaluable data on the Sun.

There are active regions on the Sun where sunspots and solar flares occur, like solar storms. They develop and go through a phase of throwing off energy—multiple flares—and then eventually die out. On Skylab, we were able to trace the history of these events as the Sun rotated. We didn't see them, of course, when they were on the Sun's back side. We could follow their history in X-ray and ultraviolet detection, as well as what was visible from the ground, to get a really good understanding of them. We also discovered something that surprised us when we looked at the Sun via X-rays. We found that there were not just the active regions or sunspot areas of high intensity that we knew of, there were also very small bright points all over—mini active regions. When you looked at it through X-ray, it was like the Sun had the measles. . . . We had to rewrite the textbooks after Skylab because there was an awful lot of new data that really changed our image of the Sun.

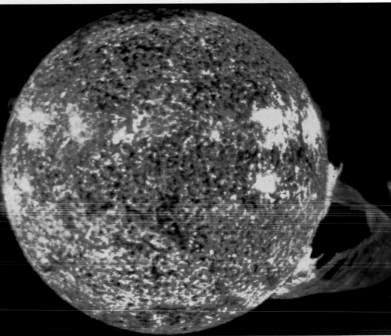

Above

An ultraviolet image of the Sun taken during the Skylab 4 mission caught one of the largest solar flares ever recorded, spanning a distance of 365,000 miles. Detailed study of the Sun was one of Skylab's main scientific achievements.

Left

Skylab 3 crew member Owen Garriott helps to install a twin-pole solar shade, a follow-up protection measure to the parasol-type sunshield that was deployed during Skylab 2.

THE APOLLO-SOYUZ TEST PROJECT

The historic meeting of three astronauts and two cosmonauts in space on July 17, 1975, was almost an anticlimax compared to the years of intensive work it took to get the crew members there in the first international manned spaceflight.

The joint mission between the Cold War opponents was the result of a formal agreement forged in 1972 following numerous discussions that began in 1969 between NASA administrator Thomas O. Paine and Mstislav V. Keldysh, the president of the prestigious Soviet Academy of Sciences.

On July 15, 1975, Aleksei A. Leonov and Valeriy N. Kubasov blasted off from the Tyuratam cosmodrome in Soyuz 19, the primary vehicle used for cosmonaut flight. Several hours later, Thomas P. Stafford, Vance D. Brand, and Donald K. "Deke" Slayton, strapped into an Apollo command module, were lofted into orbit by a Saturn IB. Both craft made the first of two historic dockings on the morning of July 17. A few hours later, Stafford and Leonov met in the specially constructed docking module that served as an airlock and transfer corridor. With applause from the two control centers in the background, the two commanders shook hands, greeted each other, and posed for pictures. During the next two days, the five spacefarers visited each others' spacecraft, ate together, exchanged symbolic items, and conducted some biological and space-science experiments.

With a "farewell" and "*dosvidanya,*" the two craft separated late on the morning of July 19. The Soviets returned home about forty-three hours later while the Americans remained in space an additional six days.

▼

The cosmonauts were more proficient in English than our guys were in Russian. Tom Stafford had a way of speaking Russian that we called "Roos-ton," which was a combination of Russian and Houston.

Glynn S. Lunney, program manager

Deke Slayton, one of the original Mercury astronauts, and Aleksei Leonov, the first human to walk in space, mug for the camera inside the Soyuz module, July 1975. Slayton had waited sixteen years to fly in space because of an earlier-diagnosed heart condition.

Thomas P. Stafford, commander

[Leonov's] a very outgoing guy, and he's like a brother to me now. You know, [the Russians] were supposed to be atheists. Once, after we got to know each other, we were speaking without an interpreter and having dinner one night in a restaurant in Moscow. . . . He told me about when he did his first spacewalk *Voskhod 2* and how the suit had ballooned. He hardly made it back in. He said [in Russian], "Thank God I got in." But he's supposed to be an atheist.

Vance D. Brand, command module pilot

Leonov and Kubasov were different from Americans in that they had different traditions and customs, but they were people, just like other people we know. It was a little hard to put a stereotype on a Russian after you got to know him. . . . Basically, it was a very enriching, human experience.

American and Russian crews share a meal during a training session in Houston months before their flight. From left: Deke Slayton, Aleksei Leonov, Tom Stafford, and Valeriy Kubasov.

Glynn S. Lunney, program manager

It dawned on us early on that when the Russians made a commitment "at the highest levels"—that was a favorite expression of theirs—to ASTP, the engineers knew they were on the hook to deliver results or face possible penalties if they didn't. Our system was different. If I had a problem with the Soviets, I could take it to my superiors in Washington and they would have supported me. Our Soviet counterparts were just given orders and there wasn't any recourse. They just had to get it done. As a result, we could push the hell out of them. We could say, "Look, if you don't do this, we're going to cancel this thing." That would send them up the wall.

I used this tactic when the Soviets refused to provide us with information on the Soyuz accident, in which three cosmonauts expired. NASA had given the Soviets a detailed account of what went wrong on Apollo 13, and we insisted on learning what went wrong on Soyuz. I would have taken it to any level in our government and I would have been supported. But they had a terrible time going back up the line to get permission to talk about it.

Richard H. Truly, capcom

Russian and American crew members took turns visiting each other's space facilities and conducting joint training exercises. In April 1975, Truly accompanied astronauts Thomas Stafford, Deke Slayton, and Vance D. Brand on their trip to Russia for final preparations.

They took us down to the launch site, which the United States called Tyuratam. The Soviets call it Baikonur [cosmodrome]. You've got to remember that there was huge intelligence interest on the part of the two nations about knowing what each other's space programs were doing. They landed and took off in the middle of the night so we couldn't see much. But we were the first Westerners other than Charles de Gaulle to set foot on the launch site. They were very open when it came to Apollo-Soyuz and totally closed about everything else that was going on down there.

We were supposed to leave Moscow at the very tail end of April, but they told us that if we wanted to see the May Day parade that we could stay another day. And so all of the MOL [manned orbiting laboratory] guys who had been on the military program, which was [Robert L.] Crippen, me, [Robert F.] Overmyer, and [Karol J.] Bobko, I think, we all said, "Damn right, let's stay." So we stayed and watched the May Day parade right below the big Kremlin Wall, where all the big shots were standing. There we were, active duty military in the Soviet Union in 1975, right as Saigon was falling, watching the May Day parade. It was very unusual.

Glynn S. Lunney

Lunney's oversight of ASTP involved all aspects of the mission, from negotiations between the United States and Soviet Union to technical and operational details to mission coordination and hardware integration.

ASTP was much more of a cultural and political challenge than it was a technical challenge. It wasn't a giant technological step by any stretch of the imagination. But working on Apollo-Soyuz taught us about the people on the other side and also about ourselves, the society that we live in, and the things we take for granted. I learned a lot more about those things than I did about engineering. I think that went on on their side as well. At one point in the discussions, my counterpart, Professor Bushuyev, said, "Dr. Lunney, this is easy for you. If you want something done, you just call up your contractor at Rockwell and tell them to do it. They know they'll earn more money by doing it. In my country it is nowhere near as simple."

With all the myths that had been planted in their minds, the Soviets got a rude shock when they spent some time in the United States. For example, they thought that NASA personnel were from some special privileged class of people. They were certain that we were descendants of important families, the founders of capitalism in the U.S. They couldn't believe it when Tom Stafford told them that his ancestors went west with the Oklahoma land rush, and that my father worked in the coal mines of Pennsylvania during his life.

Below

A group of astronauts and their cosmonaut hosts visit Red Square in Moscow during a familiarization tour in 1974. Apollo-Soyuz was as much about breaking down cultural barriers as it was about space exploration.

Right

Aleksei Leonov inspects an access hatch to the docking module that would later join the U.S. Apollo and Soviet Soyuz spacecraft in orbit. Designing a common docking mechanism was one of the mission's great engineering challenges.

▼
I understood the immense responsibility. In the eyes of all of humanity we showed the best side of man.

Aleksei A. Leonov, commander, Soyuz-19

Thomas P. Stafford, commander

In the old days, you know, Stalin just hated the British, he hated Americans, so the only English taught was in the universities, and not much of it then. But within a couple of years after the death of Stalin, English started to be the major language taught in the schools, so the very youngest cosmonauts could speak it very well. . . . So we started out ahead of them, and then by Christmas of '73 we were over there. Each one of them had a private English professor with them all the time and they were skunking us. I knew that I had to speak Russian as well as [Leonov] spoke English when we opened that hatch. So I called back to Chris Kraft, said, "Chris, we got a real problem here." And I talked to Glynn Lunney, who was also the program director. I said, "I need at least four professors full time. We need them from early morning till late at night, no union rules, Saturday, Sunday, if I'm going to make this a success." He said, "We'll do it, Tom." So we got four profs in the office, and they were with us just about day and night.

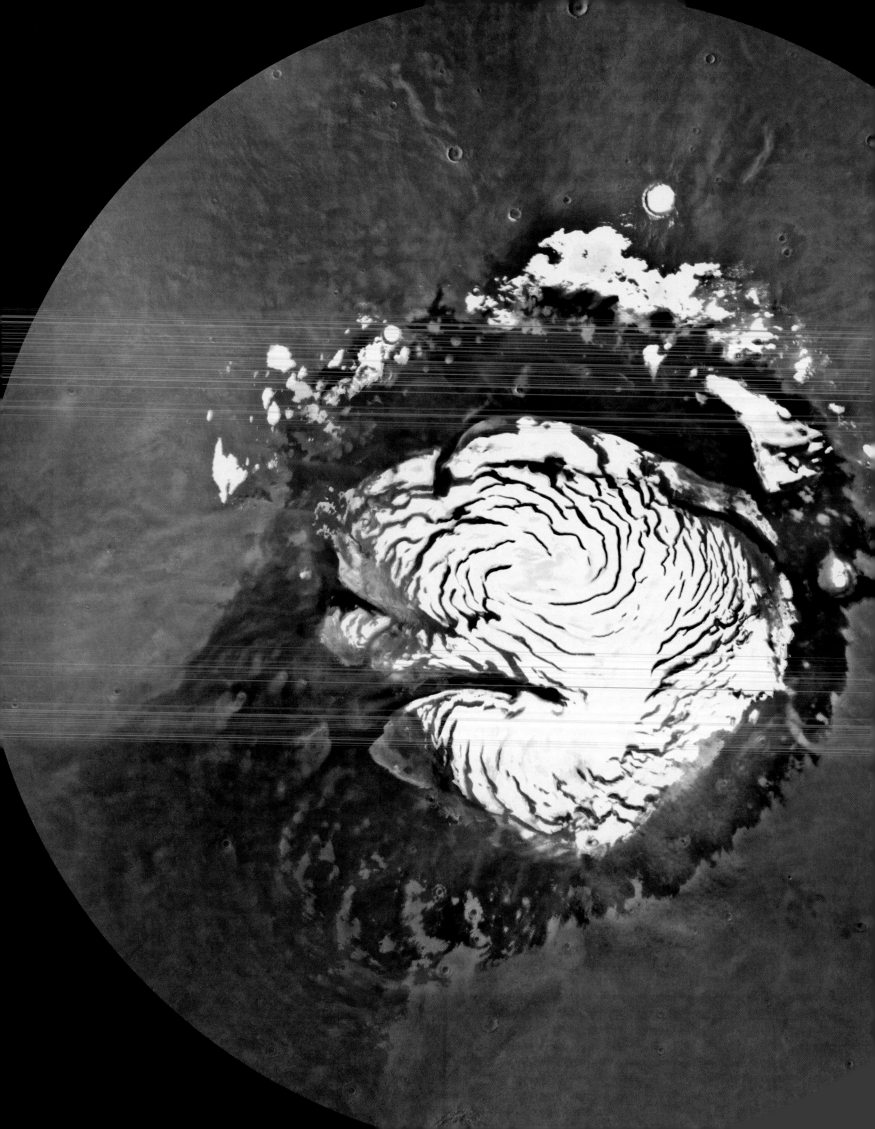

THE ALLURE OF
THE RED PLANET

Mars, a solitary ruby set in a black veil of diamonds, has captivated people throughout recorded history and undoubtedly back to the time when they first took in the grandeur of the night sky. Mars, whose red glow suggests heat, blood, and other life forces, and whose very name stands for passion and conflict, has been a persistent enigma for millennia. Mars and Earth, the red planet and the blue, are fated to sail around the Sun as intimate neighbors in the vast, dark universe.

Following Giovanni Schiaparelli's announcement in 1877 that he had spotted "canali" (channels, but also canals) on Mars, a wealthy Bostonian named Percival Lowell became obsessed with the notion that Martians had dug them to move water from the their planet's wet poles to its parched and dying midsection. He built an observatory near Flagstaff, Arizona, in 1894 and spent the rest of his life trying unsuccessfully to prove his theory. Four years later, H. G. Wells published a hair-raising novel, *The War of the Worlds*, in which Lowell's bright but beleaguered Martians invaded Earth with the intention of becoming interplanetary colonists.

Fiction aside, the possibility of life elsewhere in the solar system has intrigued humankind for centuries, and Mars seemed to be a likely place for it. While no scientist at the dawn of the space age seriously believed that there were canals and little green men on Mars, there remained the possibility of vegetation or microbial life. But no earthbound telescope could bring the Martian surface close enough to find out for sure. Scientists and engineers knew that robots would have to be sent to Mars to settle the

The northern polar cap on Mars is revealed in this mosaic of images taken by the Viking orbiter. The grooved ice cap is surrounded by flat plains and large dunefields.

▼

Mariner 9's close-up

portrait of Mars not only showed the planet to be complex and interesting, but it provided enough detailed information for the next step in the exploration process: a landing.

matter. The search for life on Mars therefore became the most powerful incentive to go there.

The goal was to come up with a plausible exploration strategy that would, in stages, lead to ever-increasing information sent back by robots and then to an eventual expedition and colonization by humans. As was true with the Moon, the goal would be realized by each succeeding mission sending back better data: Flybys would pave the way for orbiters, orbiters would pick out terrain suitable for landers, and landers would photograph their surroundings and conduct experiments, including a search for life.

But the road to the Red Planet was strewn with obstacles. From a political standpoint, the constituency for Mars exploration was relatively small, which had a direct influence on appropriations. (Many Americans were deeply curious about Mars, but relatively few wanted to pay tax money to find out about the place.)

Technologically, there were challenges everywhere. Spacecraft would have to be designed to carry enough scientific instruments to make the trip worthwhile, which would make them relatively heavy, requiring powerful and expensive rockets to get them on their way. They would be completely on their own, very far from home, and would therefore depend on solar panels for electricity, excellent navigation systems, and first-rate two-way communication systems. In addition, the spacecraft would have to be steered precisely to their target after being launched during the one "window" every twenty-six months when Earth and Mars are best positioned for such missions.

As early as 1958, the National Academy of Sciences' Space Science Board had an ad-hoc committee that recommended sending probes to both Venus and Mars. That year, two Mariner-class flybys to Mars were proposed for as early as 1960, but the mission fell victim to budget cuts and development problems with the powerful Centaur upper stage that was supposed to give them their final shove. Almost $15 million was finally appropriated for Mariner in 1961. The deep-space probes were designed and built at the Jet Propulsion Laboratory in Pasadena, California. Mariner 1, destined for Venus, went to the bottom of the Atlantic Ocean instead when its booster went off course on July 22, 1962. Its stablemate, Mariner 2, became the first visitor from Earth to reach another planet when it skimmed by Venus later that year. Mariner 3, bound for Mars in November 1964, failed to fulfill its mission when its protective shroud failed to eject as it entered space, thus preventing its energy-gathering solar panels from deploying. The probe's twin, Mariner 4, reached Mars in July 1965, sending back the first close-up pictures of the planet.

Mariners 6 and 7 skimmed past the planet in February and March 1969 and returned two hundred detailed television pictures. Robert B. Leighton, the eminent Caltech physicist, knocked down the notion that Mars was roughly similar to Earth in that it had polar caps, seasons, and a twenty-four-hour day. Mars, he reported, was "like Mars." That is, it was uniquely Martian. Mariner imagery showed extensive cratering, which was reminiscent of the Moon. Norman H. Horowitz, the Caltech biologist, saw nothing in the data that indicated life.

But during the next window, in May 1971, Mariner 9 swung into orbit around Mars rather than simply flying by it. It returned extraordinary imagery that changed forever our view of the Red Planet.

Circling the planet repeatedly increased the science return enormously. The robotic explorer globally mapped the planet in spectacular detail. But other important data on infrared and ultraviolet radiation, magnetic fields, charged particle levels, and more were collected to provide the fullest possible description of the planet.

Mariner 9's close-up portrait of Mars not only showed the planet to be complex and interesting, but it provided enough detailed information for the next step in the exploration process: a landing.

The search for life on Mars entered a new phase in the summer of 1976 when Vikings 1 and 2 landed on the planet's surface. Both searched for signs of life, past and present, but the result was inconclusive. Despite that, the Vikings sent so much data back to Earth that their summaries alone filled a nearly fifteen-hundred-page book called, simply, *Mars.*

One thing was certain. The enigmatic red dot in the night sky would draw intrigued earthlings back to it.

Mariner 4 views the Red Planet up close, July 1965. The first spacecraft to explore Mars returned just twenty-two pictures, but they were enough to show a dry, cratered landscape devoid of life.

MARS

MARINERS 4 & 9

Four lightweight Mariner-class spacecraft reached Mars between 1965 and 1971. Of them, the voyages of the first and the last demonstrated how planetary exploration builds on successive missions.

Mariner 4, launched in November 1964, flew by the Red Planet on July 14, 1965. It reported that the Martian atmosphere was thinner than expected and was heavy with carbon dioxide. Through a specially designed television camera, it also sent back twenty-one complete images and one partial image, all of which were fuzzy by today's standards. The pictures were both exciting and disappointing: exciting because they were the first close-ups of the enigmatic planet; disappointing because they indicated that Mars was a simple, crater-pocked orb like the Moon.

Wrong! Mariner 4 had photographed only one percent of Mars as it sped by. Mariner 9 literally hung around. Over the course of a year, it sent home 7,329 detailed television images of the entire planetary surface, along with other data.

The imagery was stunning. "Four huge, dark mountains poked through the dust-laden atmosphere," Clark R. Chapman, a planetary scientist, would later recall. They were extinct volcanoes, not meteor-impact craters, which proved that Mars has had an active inner life (geologically speaking). The largest, Olympus Mons, is the size of Colorado. In addition, the ecstatic scientist reported that there were vast canyons that implied evidence of geological forces similar to those on Earth. Some walls suggested water erosion, meaning the possibility of life.

Mariner 9 proved beyond doubt that Mars was worth a closer look, and it provided the maps that would make the Viking landings possible.

▼

The biologists were disappointed with Mariner 4, but it was a triumph for me as a geologist because we discovered that Mars was a different place than we thought. What else could you want? I never had the expectation—and I don't think anybody on our imaging team did—of seeing lifelike features.

Bruce Murray, professor, California Insititute of Technology

Mariner 9's images of ancient river channels turned scientists' conception of Mars completely around. By finding evidence of running water in the planet's past, the 1969 mission revived the hope of discovering fossil life.

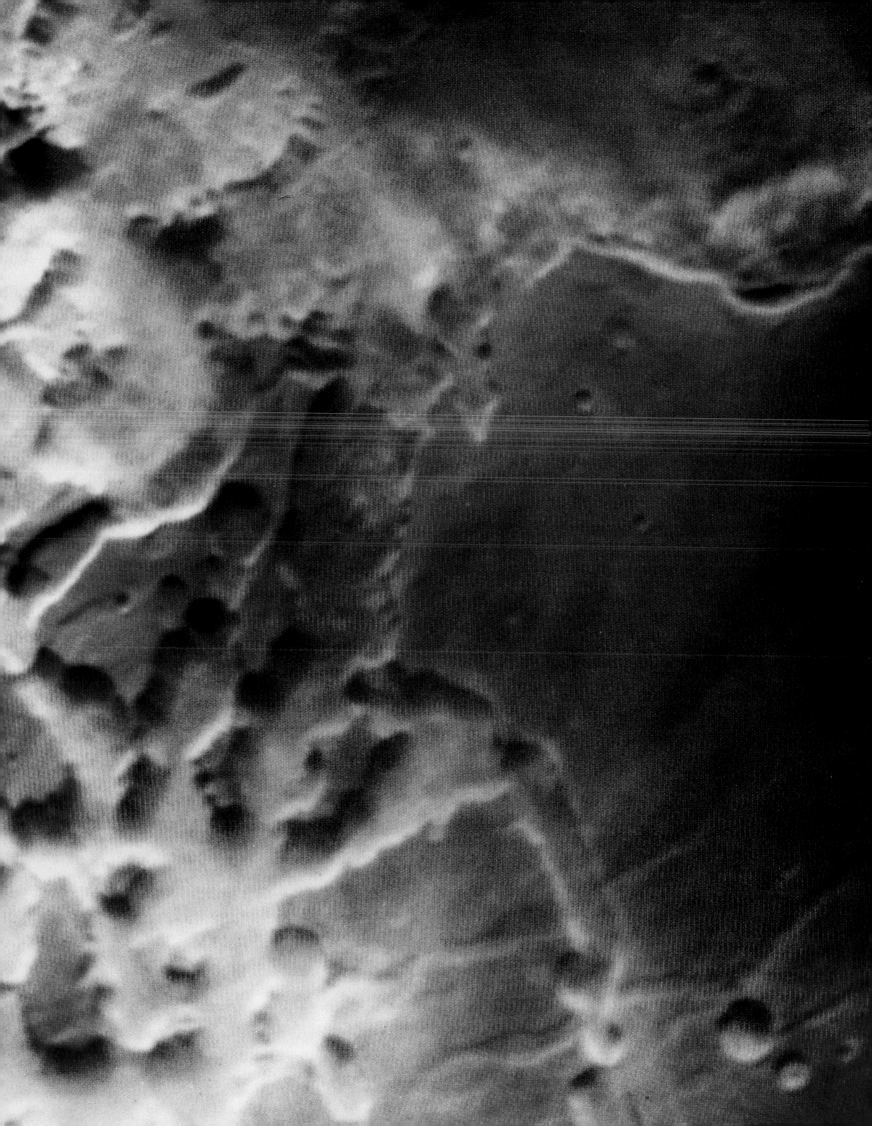

MARINERS 4 & 9

Bruce Murray, professor, California Institute of Technology

Murray was a key participant in the development and interpretation of the television images that were captured by the cameras aboard Mariner 4, Mariners 6 and 7, and Mariner 9.

Nobody had created a digital TV camera of this kind before Mariner 4. We did it in space before it was done on the earth, so the whole technology was new. We had to use a TV sensor system that could store a picture for a long time—eighty-four seconds—on the vidicon surface because we didn't have a tape recorder that we could run fast enough to take the image off.

That camera was humanity's eye in space. Like it or not, that's what we had to use. We had done simulations in the lab, but we didn't know what the first pictures of Mars were going to look like. There were no test pictures taken in flight.

The first things we saw were images that showed no detail at all. There was a signal there, but it looked like the back of a white envelope. Part of this was due to the fact that the lighting was bad. If you view the Moon or Mars with the Sun behind you, you'll see virtually no topographic detail. And if the planet has an atmosphere, as Mars does, it's even worse.

Mariner 4 collected over five million bits of data, which were stored on a tape recorder. At the painfully slow rate of 8.33 bits per second, it took over a week for all of the data to return to Earth.

It took eight hours for each image frame to come in, and they came by teletype from the Goldstone antenna in California [one of three deep-space stations]. There was no immediate full picture, just strips of numbers ranging from 1 to 128. There were twenty-one complete images in total.

This was the first time anybody had dealt with these images. When we got them, they had aberrations due to scattered light. (We believed, but never proved, that dust particles in the spacecraft had migrated around and ended up on the camera mirror.) We had to work like mad to try to invent digital image processing. We worked day and night for three days.

In the meantime, for the TV networks (there were only three at the time), this was all a big deal—the first look at Mars. The camera crews were all parked at the Jet Propulsion Laboratory. They were raising hell because we were not showing them our pictures. We're sitting on the data! We're hiding it! We're using public money and not showing the results! They were getting pretty forceful.

We didn't say, "Look, we don't know what's going on. We're trying!" Gradually we figured out how to do the image processing and began to get pictures that did show markings on the Martian surface. My moment of truth—and I have this still in my office—was to take two of these overlapping prints and slide them together. And the markings matched! I knew these were features on Mars, because they appeared in two different frames. We were finally seeing the surface. . . .

Jules Bergman, who was the ABC-TV science reporter, was at the press conference when Bob Leighton, head of the imaging team, was giving his pitch. Leighton then asked for questions. Bergman got the microphone and asked, "Doctor, what do you see down there?" Leighton was supposed to say something newsworthy that could get quoted, but he wasn't that kind of person. "Dirt," he replied. Bergman never forgave him.

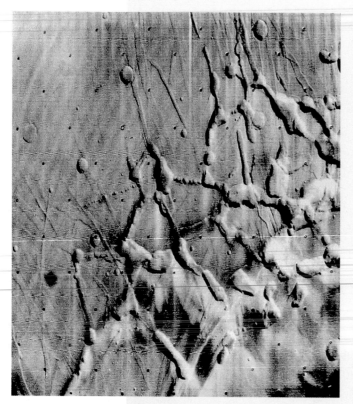

Topping the list of spectacles revealed by Mariner 9 was the vast canyon system known as Mariner Valley, or Valles Marineris. The "labyrinth" at the canyon's western end is made of dozens of intricate grooves and crater chains.

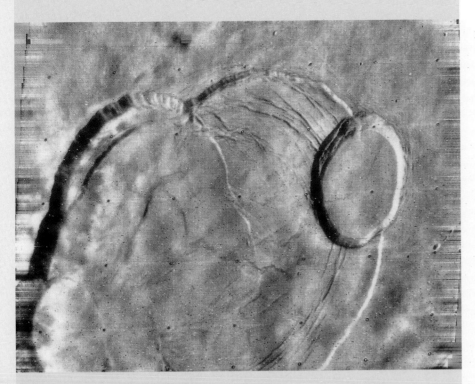

Frank Jordan, orbit determination leader, Mariner 9

Orbit determination involves calculating the trajectory necessary for a spacecraft to achieve a desired orbit. After establishing Mariner 9's orbit, Jordan's team discovered, before the cameras did, the volcanic area in Tharsis, which includes Olympus Mons, the solar system's largest volcano.

I had a team of about eight or nine people. We calculated the transit-to-Mars trajectory and the Mars-approach trajectory. I personally calculated the first orbit around Mars. That was in November 1971.

I still remember the night when we determined the first orbit, a twelve-hour orbit. The maneuver team had secured it with a retro-maneuver, and we calculated it from the Doppler data. We established within about a day that there was a great big "mound" of gravity in what we now know as the Tharsis bulge area. And we "saw" this before it was visually seen, because there was a surface dust storm occurring. We basically knew the gravity shape of Mars weeks before anybody saw craters and relief. The cameras just saw nothing of the Martian surface.

About eight hours into Mariner 9's first twelve-hour orbit, I knew we had something that was not like anything on Earth. I didn't know where it was or what it was. Within two orbits, maybe half a day later, we had a gravity map that showed a big mound. When the pictures came out, it was no surprise. In that volcanic region, all of the material that had been coming up has shaped the gravity field of the planet. There had been extensive lava flows for centuries, and basically whatever had been inside Mars came out and piled up on that side of the planet, creating an egg shape, physically. And gravitationally, it was an enormous anomaly. Every time the spacecraft passed through a periapsis [the closest distance a body in orbit reaches to another mass], there was a big extra signal boom when the spacecraft got close to the surface.

I was thirty-one years old at the time. I remember that my wife and I brought our first son home from the hospital the morning of that encounter. We named him Marc for "Mars."

Stamatios Krimigis, investigator, Mariner 4

In addition to doing surface studies, Mariner 4 also performed magnetic field and particle measurements. Krimigis refers to "dynamos," which are how rotating planets generate magnetic fields.

Our instrument on Mariner 4 was designed to search for the equivalent of Van Allen belts in the vicinity of Mars. In those days everybody thought that since Mars was not much smaller than Earth, and it turned on its axis every twenty-five hours, and dynamos were "known" to be driven by planetary rotation, we would detect a big magnetosphere, radiation belts, and auroras on Mars.

We flew by at around ten thousand kilometers [approximately six thousand miles] from the planet, which is not that far. If you approach that closely to Earth, your detectors are going to light up like a Christmas tree. Basically, we saw nothing. It was a great puzzle. . . . It wasn't until Mars Global Surveyor that definitive proof of magnetism on Mars was found. It turned out to be remnant magnetism indicating an area of some past tectonic activity.

Above

A Mariner 9 close-up of the collapsed crater atop the solar system's largest volcano, Olympus Mons. The mountain towers 18 miles above the surface of Mars.

Mariner 4 carried a television camera and six science instruments on its trip to Mars. The communications technology of the day allowed for only a handful of pictures, beamed back to Earth at the agonizingly slow rate of eight bits per second.

VIKINGS 1 & 2

With Mariner 9 having led the way with the first orbit of another planet, NASA dispatched two new explorers to Mars in late summer of 1975.

The Vikings were designed as heavy, highly sophisticated combination orbiter-landers. The orbiters were to map the planet in unprecedented detail, collect scientific data, including seasonal weather information, and relay data collected by the landers back to Earth. The landers would photograph the terrain around them, perform atmospheric and other sampling tests, and most important, search for the existence of past or present life.

Both landers carried three experiments that were each built into a microbiology instrument about the size of a one-gallon milk container. These miniature laboratories would analyze soil samples scratched from the Martian surface by the lander's mechanical claw.

Two of the experiments would search for microorganisms by exposing the soil to organic nutrients, then detect any emission of gas as a waste product. The third, which belonged to Caltech biologist Norman H. Horowitz, would determine whether microorganisms created organic compounds out of radioactively labeled carbon dioxide and carbon monoxide gases.

Viking 1 landed on the Chryse Plains in the northern hemisphere of Mars on July 20, 1976. Viking 2 set down on the Utopian Plains on September 3, farther north and on the opposite side of the planet. The first pictures of the ocher, boulder-strewn Martian surface and of the salmon sky above brought cheers from mission scientists and engineers who savored the historic moment. The science return was extraordinary. But the search for life was inconclusive. On that score alone, Mars remained a tantalizing enigma.

▼

I thought the conditions were not very suitable for finding anything microscopic, and that turned out to be the case. But I was all for trying. You don't do experiments to find out what you know, you do experiments to find out what you don't know.

Philip Morrison, professor of physics, Massachusetts Institute of Technology

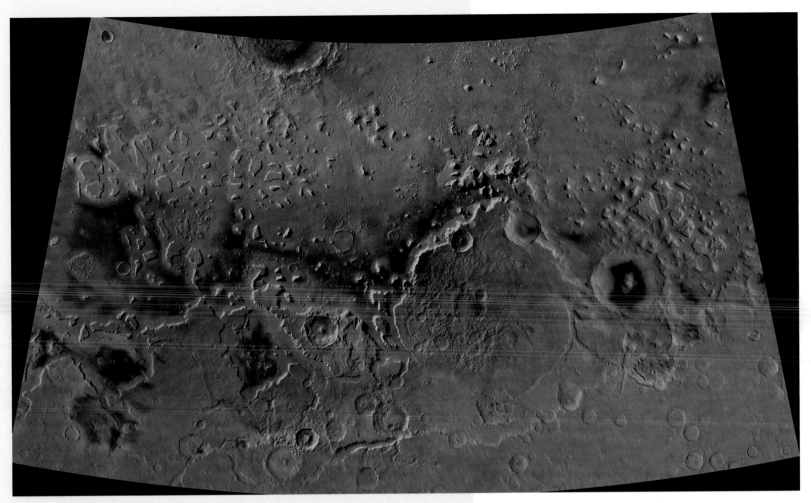

Gerald A. Soffen, project scientist

I think most people felt before the mission that we weren't going to get any definitive sign of life. But there was always a chance we could. So when the gas chromatograph/mass spectrometer result came back [showing no organic material in the Martian soil] it was the biggest surprise of all. It is still, to this day. I would have bet my life on it. I was sure there was organic material on Mars. There was supposed to be. And of course, there wasn't, at least not at the two sites we went to.

The complexity of Martian geology is apparent in this composite of Viking orbital images showing part of the planet's Ismenius Lacus region. Among the landforms in this scene are craters, mesas, and flat-floored valleys.

VIKINGS 1 & 2

Hans-Peter Biemann, high-school student

Biemann's father, Dr. Klaus Biemann, led the molecular analysis team that analyzed Martian soil samples for organic compounds. The then-sixteen-year-old witnessed and documented the Viking mission in his book The Vikings of '76. *This is his account of the landing while present at the Jet Propulsion Laboratory:*

As the excitement increased in Building 230 [spaceflight operations facility], I tore myself away and raced over to Building 264. Arriving in the biology area at 4:50 P.M., twenty minutes before landing, I found the group gathered there already bubbling over with excitement. . . . Gentry Lee [director of science analysis for Viking] translated the engineering data on the screen for the biologists, tracking the activities aboard the lander.

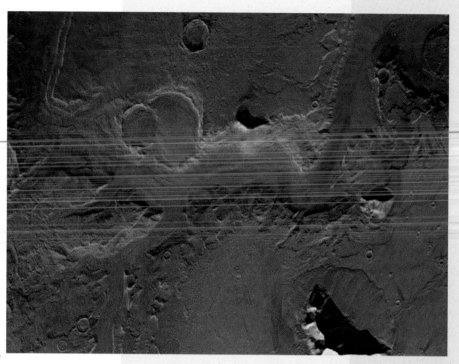

"It's going too fast!" cried Gentry a few times in fear that something had gone wrong and the lander would crash. But all was nominal. The excitement and tension had distorted his sense of time. Later, terror struck the group when the data flow halted; but that was a preplanned, normal event.

Everything was functioning beautifully. On one screen appeared the nominal descent curve. The actual data points were falling smack on the curve, indicating a flawless entry. Viking was three hundred thousand feet away from victory. . . or disaster.

Five short, tense minutes later, Gentry announced the parachute's successful opening. Only seventy-four thousand feet to go! The room throbbed with excitement. Every single person was totally caught up in the emotion. "Forty-eight hundred feet." Everyone listened to Gentry. "Twenty-six hundred feet." Viking was flying toward the orange plains of Chryse. The biologists shouted with joy at every announcement.

"After fifteen years, we're finally landing on Mars!" Gerald Soffen was consumed with wonder, awe, and disbelief. It had been so much work, and now he and the hundreds of other "Vikings" at JPL faced the moment of truth.

"Five hundred feet!" shrieked Gentry. The atmosphere was electrified.

"Two hundred feet!"

"Touchdown!" Everyone was shouting. Cheers could be heard over the Net. We knew Viking had landed, but had it survived?

There was little time to ponder that point. Ten seconds after touchdown, the bit transmission rate changed from four thousand bps [bits per second] to sixteen thousand bps. That was it. The bit rate had changed. Viking had landed safely and was ready to do its thing.

A few hundred yards away in Building 230, Rex Sjostrom, LPAG [lander performance analysis group] chief, cried, "We've got a good one!"

As the fervor continued, champagne corks popped and cigars lit up. The thrill of victory thumped in everyone's hearts.

E. Myles Standish, astronomer, Jet Propulsion Laboratory

In order to fill the navigation needs of spacecraft traveling through the solar system, sophisticated data tables, called ephemerides, are necessary to calculate the ever-changing positions of planetary and other celestial bodies. Standish has been the main planetary ephemeris specialist for JPL since the early 1970s.

The very first Venus shot [Mariner 2] was a basket case. It was really bad. The ephemerides were taken from the Naval Observatory's Blue Book, which is just a regular astronomical almanac. Previously, the Blue Book had been used simply to point a telescope at a point in the sky and look for a particular planet. As time went on, and as we've become more accurate, mission planners ask for more accurate positions. Viking, I think, required approximately a thirty- to fifty-kilometer [eighteen- to thirty-one-mile] error margin, not much more.

Above

A channel flows through it: Viking's view of the wide channel known as Reull Valles, in the southern hemisphere of Mars. Colors in the image have been computer-enhanced to show different terrain types.

Ann Druyan, writer-producer

Renowned Cornell University astronomer Dr. Carl Sagan played a leading role in NASA's Viking expedition to Mars. Druyan, his collaborator and later wife, was with him at California's Jet Propulsion Laboratory when Viking landed.

During the preparatory phase of the Viking lander design, Carl was extremely worried that the imaging system on Viking was extremely slow. It did not work the way a camera would work, but instead the way a fax machine would work: by scanning and painting in one strip of information at a time. He was really worried that if there were any type of visible life—more than a microbial kind of life—on Mars, the slow, noisily functioning cameras wouldn't catch that life. And so—and this so typically Carl—while in the American Southwest during the testing of the Viking camera, Carl went to a pet store. He bought a turtle, a lizard, and something that looked like a mouse. I think it was a field mouse. And when no one was looking, he set them loose in front of the camera. The camera did not catch the lizard at all and it did not catch the mouse. It only caught the turtle, the slowest moving of the three, in such a way so it was still completely unrecognizable. That was exactly the way that Carl thought.

Gerald A. Soffen, project scientist

We were originally supposed to land on the Fourth of July, and it was a big deal. There was the question of what would appear in the first picture taken of Mars: Would it be the American flag on the lander, or Mars? We could take a picture of the flag and still get a picture of Mars over the edge of the spacecraft. We finally decided there was no question that the first picture had to be of the lander's foot, but the very next one had to be a picture with the flag in it. We had a big battle with the scientists, who said, "We're not here to take pictures of the flag." I was in the middle trying to make peace. We compromised—we took a picture of the flag, but not in the first two shots.

Frost-covered rocks at the Viking 2 landing site. Although the planet's large stores of water have long since vanished, some moisture is still exchanged between the atmosphere and the surface. The ice shown here is barely a thousandth of an inch thick.

VIKINGS 1 & 2

Gerald A. Soffen, project scientist

When the first pictures from the orbiter came down, there were probably ten of us looking at the images, which were each about the size of a postcard. We were all huddled around one picture. And I remember Hal Mazursky, a geologist on the imaging team, saying, "There's no way we're going to land on that." The pictures looked just like the badlands of the Dakotas. So we went out to supper to try to figure out what the heck we were going to do. Of course, we didn't land on the day we were supposed to.

I just prayed that we'd get the first lander down somehow, then we could locate where the second one was going to go. The second lander was a month behind the first one. But I worried that we didn't have enough time.... We had everybody trying to figure out what to do. We had Ph.D.'s coloring in craters, doing crater counts. Everything stopped to try to come to grips with the problem, and there was meeting after meeting. It was the first time I brought a sleeping bag into work, and I used it many, many times.

Michael C. Malin, member of the technical staff, Jet Propulsion Laboratory

I think anyone who said the first Viking pictures were what they were expecting was probably lying. I don't think we had the slightest idea what Mars was going to look like. I was quite thrilled with what it *did* look like. The first panorama was at fairly low resolution and, boy, we thought we saw little stream channels and gullies and all sorts of stuff. It was really quite strange. And ultimately we got some exquisite mosaics from the Viking cameras. It wasn't as diverse a scene as some place on Earth might have been, but you know what? It was Mars. And that was the impressive thing. It was what *Mars* looked like.

Panorama of the Viking 1 landing site. The boom extending upward just right of center holds instruments for studying the Martian atmosphere. The American flag was both a symbol of national pride and a calibration target for the cameras.

I never dreamed in my wildest imagination that Viking would last as long as it did. That was the biggest surprise of all. We were beating the Martin Marietta company over the head with, "It's got to last overnight. And if it lasts thirty days, which is our goal, we'll give you the biggest reward you ever saw." And of course, it lasted five years.

Gerald A. Soffen

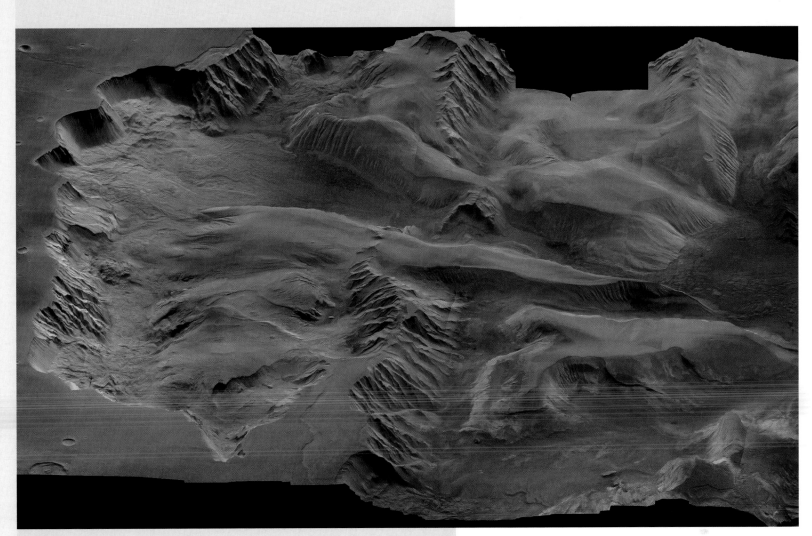

Norman H. Horowitz, chief, bioscience team

The Gas Exchange Experiment (GEX) was designed to measure gaseous products from a sample exposed to a humid atmosphere or treated with a distilled water solution. Atmospheric change occurring in a sealed sample container could indicate microbial metabolic activity.

I thought at the time that the GEX was a lousy experiment because we knew there was no liquid water on Mars. But it turned out to be terribly interesting because it was the first to reveal the chemical reactivity of the Martian surface. So, it didn't find what it was looking for—it couldn't have found life, in my opinion, because I don't think that Martian organisms, if they exist, would know how to handle all that water—but it looked for a chemically reactive surface and found it. It was the first evidence from the products that were formed that the Martian surface contains peroxides and superoxides—products of oxidation.

Gerald A. Soffen

Before the landing we held a pool to guess what the surface of Mars would look like. We collected a lot of pictures from a lot of scientists. There were probably fifty people in the pool. We collected a nickel apiece, or something like that. . . . The guy who won wasn't a scientist. He had entered a picture of a place near where he lived in the scablands of Washington state. And that's of course, exactly what Mars does look like. He had just Xeroxed something out of a travel book, and it looked very much like the landing site.

We all laughed and said, "Whatever happened to all of our famous geologists?"

Years after the spacecraft went silent, scientists still use Viking data to explore the Martian surface. This 3-D view of Valles Marineris was assembled in 1991 from high-resolution black-and-white images and low-resolution color images taken from Mars orbit.

SHUTTLING TO SPACE

T he notion of a station to keep people in space all the time depends squarely on a reusable ferry—not Apollo-type "cannonballs"—from which astronauts could construct it, change crews, and keep it supplied. The relationship between ferries—or shuttles, as they came to be called—and stations is a fundamental one, proposed by such visionaries as Wernher von Braun, who described a shuttle in detail on the pages of *Collier's* magazine in 1952 and Arthur C. Clarke, who featured shuttle-to-space-station use in his 1968 novel, *2001: A Space Odyssey.*

NASA's shuttle was conceived in 1969 in conjunction with plans for a space station, even as the end of Apollo loomed on the distant horizon. The agency looked beyond the Moon program and came up with what it called the "next logical step." That is, a space shuttle—officially called the Space Transportation System (STS)—and then a station. But the idea was seriously oversold to Congress. NASA claimed that launch costs would be reduced from those of the Apollo program by a factor of ten and that there would be fifty-five launches a year. The agency's original shuttle design called for two reusable vehicles, one a lifter, the other an orbiter. Ultimately, Congress approved a less-expensive, more-complicated, and partly expendable version.

Space shuttle
Columbia awaits its first trip into space in 1981. The shuttle revolutionized space travel. Not only could it land on a runway, it could return to orbit—the world's first reusable spacecraft.

▼

Five hundred and fifty-one

people went to space in shuttles during their first nineteen years of operation, logging 802 days of flight time, some of it to begin construction of the International Space Station.

The final design of the STS consisted of four primary parts: a winged orbiter (or orbiting vehicle) with three main shuttle engines; an external tank containing hundreds of thousands of gallons of liquid hydrogen and liquid oxygen housed separately and used to feed the main engines; and two "strap-on" solid-rocket boosters (SRBs), the largest solid propellant motors ever built, which are attached to the external tank and supplement the main engines.

During launch, hydrogen is pumped into the main engines, which burn for roughly eight minutes. The solid-rocket boosters assist the main propulsion system in getting the shuttle off the pad and are jettisoned once their fuel is expended, roughly two minutes after launch. They float down to sea under parachutes to be recovered, refurbished, and flown again. The main shuttle engines shut down just before the craft is inserted into orbit, after which the external tank separates from the orbiter, breaks up, and falls into the ocean.

The orbiter spaceship owes its design to two airplanes. One, the North American X-15, was a sleek black rocket plane that established an altitude record of 67 miles in 1963 and a speed record of 4,520 miles an hour four years later. It provided vital knowledge of how planes perform at very high speeds and altitudes and how they are controlled: knowledge basic to shuttle design and operation. The other plane was called a lifting body. A series of these, all stubby, wingless, and shaped roughly like half a pumpkin seed cut the long way, were used to see how spacecraft like orbiters would react to reentering the atmosphere at high speed.

Through the late 1970s the first orbiter, *Enterprise*, a so-called boilerplate version, underwent tests including an approach and landing program. This and other testing paved the way for *Columbia* to make the first shuttle trip to space on April 12, 1981, with John W. Young and Robert L. Crippen at the controls.

NASA's orbiter fleet eventually grew to include *Atlantis*, *Discovery*, and *Endeavour*, the last of which replaced *Challenger*, which was destroyed in a 1986 explosion that killed its crew. *Enterprise*, not equipped for spaceflight, eventually became the property of the Smithsonian Institution.

All orbiters are aluminum structures protected by heat-resistant materials on all sides. Their underwings, bellies, and tail leading edges are covered with tiles that can withstand temperatures of twenty-three hundred degrees Fahrenheit. Each orbiter has a payload capacity of about sixty thousand pounds, easily large enough to carry one or more heavy satellites and Spacelab, the self-contained science experimentation laboratory built by the European Space Agency. Payloads can be moved out of the payload bay either by the Canadian-built remote manipulator arm or by astronauts doing extravehicular activity with manned maneuvering units (MMUs) that are strapped on like powered backpacks.

Orbiters are designed to be flown by a pilot and a mission commander. Mission specialists, who can be scientists or engineers, conduct experiments and do other specialized tasks. Payload specialists manage the payloads. The normal crew complement is seven, but a total of ten persons can be carried in emergency situations. Landings are made either at the Kennedy Space Center (KSC) or at Edwards Air Force Base in California. Orbiters that land in California are attached to the top of 747s and flown back to the KSC. Once back on the ground, the orbiter undergoes maintenance and inspection procedures at the Orbiter Processing Facility.

For all of its birth pains, the STS became a transition vehicle from the cramped capsules to the International Space Station. With a flight duration of two weeks or more and ample space for habitation, shuttles permit astronauts to live comfortably, rather than being strapped to a seat. Food squeezed out of tubes has given way to real food such as smoked turkey and mushroom soup, while body waste now goes into a toilet of sorts for processing. An orbiter like *Columbia* is to an Apollo capsule what a corporate jet is to a single-engine plane.

Five hundred and fifty-one people went to space in shuttles during their first nineteen years of operation, logging 802 days of flight time, some of it to begin construction of the International Space Station.

This permanent fleet of orbiters allowed the astronaut corps to expand (while the ranks of the fighter jocks thinned) to include a large and diverse group of men and women, including participants from other countries, some of which had relatively small space programs of their own or none at all. Sally K. Ride became the first American woman in space when she flew on *Challenger* in June 1983 on the seventh shuttle flight. That same year, Guion S. Bluford Jr. became the first African American astronaut in space.

The transition from sending astronauts on quick flights to the Moon to keeping them in space for longer missions marked the beginning of a true space culture, taking us closer to permanent occupation in space.

A space-imaging radar (SIR-C) antenna sits in the cargo bay of space shuttle *Endeavour* with a view of the Pacific Ocean in the background. Instruments like this one have returned unprecedented views of Earth from orbit.

COLUMBIA ORBITS EARTH

Both U.S. and Soviet space program managers delighted in marking famous occasions with important missions. Perhaps that is why the first shuttle to take to space, *Columbia*, did so on April 12, 1981: the twentieth anniversary of Yuri Gagarin's historic flight.

Columbia's inaugural flight, STS-1, opened the shuttle era and was the first U.S. human spaceflight to employ solid rockets. It was also the first of four intensive Orbital Flight Test missions that would determine whether the entire Space Transportation System and the orbiters themselves functioned as they should under the severe stress of liftoff, flight, and landing. During their two-day, thirty-six-orbit flight, astronauts John W. Young and Robert L. Crippen tested the system in detail, including the cargo-bay doors and the maneuvering systems that change orbits and attitude control. The only payload onboard was a Development Flight Instrumentation (DFI) package that contained sensors to record the orbiter's stresses and performance.

Postflight inspection of *Columbia* showed a loss of 16 of more than 30,000 heat-absorbing ceramic tiles protecting its belly. An additional 148 tiles were damaged. It was determined that the tens of thousands of gallons of water that were poured into the flame pit and onto the launch platform to minimize the otherwise destructive flames from the solid rockets were inadequate. Subsequent modification eliminated the problem.

But such evaluations are what flight-testing is all about. Most basically, *Columbia*'s maiden flight was considered an unmitigated success and proved that the shuttle concept worked.

▼

The ride on the first stage with the solids is like driving a pickup truck down an old washboard country road.

Robert L. Crippen, pilot

Liftoff of STS-1, April 12, 1981. For the first time, astronauts were onboard an untested spaceship making its initial flight.

COLUMBIA

Steven A. Hawley, astronaut

Prior to STS-1, Hawley served as a simulator pilot for software checkout at the Shuttle Avionics Integration Laboratory .

On April 10, we had a software problem, which caused us to scrub the launch. It was our job at the facility to see if we could reproduce and understand that failure. We did. So, some of that work allowed us to support the successful launch that happened two days later.

The failure had to do with a quirk that could occasionally happen when the software was initially loaded. So, on the attempted launch day, the manifestation of the failure did not happen until, as I recall, inside T-minus twenty minutes, but the problem had been there ever since the computers had been loaded days before. We just did not know it. So, what we ended up having to do for a day after the scrub was do testing where we were loading the computers to see if we could reproduce this low-probability event. This is called an initial program load and involves pushing the button and throwing some switches in the cockpit. I think I did a hundred and fifty-some initial program loads in the facility, which I think is probably still a world's record among astronauts for the number of times to have loaded the computers. We were finally able to reproduce the problem. It was not a serious problem once we understood it.

Joseph P. Allen, scientist-astronaut

In 1978, Allen was assigned to the Operations Mission Development Group, where he served as a support crew member for the first orbital test flight of the space shuttle.

Here you are, given a new set of hardware to use, namely the Space Transportation System. One way to understand the challenge is, here's a new automobile that comes from the manufacturers, who have never driven it. But they've put it together, and it's engineered to be quite a fancy automobile. And there it sits, along with a prototype owner's manual written by people who have never driven it.

As an operations person, you begin to write your own operations manual. You make sure you really understand how fast this is going to go, how fast it's going to accelerate, and where you should first go with it. And you lay down the test protocols that you're going to undertake when you actually get in that "car" and start the engine. . . .

Now, STS-1 was to be the orbital test. So, we worked on how we would choreograph the actual operation of this thing: What path should it follow? What acceleration profiles should we load into the computer, such that engines are throttled or not throttled? All those sorts of things were built from nothing, from clean pieces of paper, and training plans were set up—I became involved in all of that. It was quite fun stuff. This was a machine that had never flown before. People were going to fly in it, and they were going to go from sea level to two hundred miles high in about eight minutes. They were going to go from a standing start, not moving at all, to going eighteen thousand miles an hour in those eight minutes. And that's like riding a lightening bolt, I'll tell you that.

Veteran John Young (left) and Robert Crippen preview some of their intervehicular activities prior to flight. Young had flown the first Gemini flight in 1965 and visited the Moon twice.

Astronaut John Young is caught in a pensive moment

in Henry Casselli's watercolor *When Thoughts Turn Inward.*

Henry Casselli, artist, NASA Art Program

In 1963, NASA began commissioning artists to document its space exploration efforts. Casselli was present in the White Room with astronauts John Young and Robert Crippen just prior to launch.

When Thoughts Turn Inward came out of the actual day of the STS-1 launch, a second attempt after a three-day delay. They had gotten a little bit ahead of the clock so there were a few minutes for Young and Crippen to just sit there without a lot of activity, questions, or anything. The entire room became quiet. For Young, it seemed as though his thoughts had turned inward, which was the only way I can describe it. He could have been thinking on any one of a thousand things, but it was obvious it was not the time to engage in conversation with him. I just sat quietly on the floor between him and Crippen, looked up, and worked on sketches and a very loose watercolor. To be honest, I was fearful that Young would raise his eyes and look at me and I would just know, without a word, that I was invading a very private space. But that never happened because by that time he and I had gotten to know each other well enough and he knew why I was there. I was totally accepted, but also fortunate to be at the right place at the right time.

Robert L. Crippen, pilot

The shuttle is a very complex vehicle, and I fully anticipated that we were going to have to do the countdown several times before we might actually get to a liftoff. We had tried it on April 10 and had a computer glitch, but got back in on the twelfth. I wasn't overly confident that we were going to fly. It was only when the count got inside a minute that I looked over at John and said, "I really think we're gonna do it." At that point, I started to get pretty excited. My heart rate went up to something like one hundred thirty, and John's stayed down at ninety. You don't work on something that long without—when the opportunity finally gets there—getting pretty excited.

When the main engines lit off, I could monitor those on my instruments as well as feel the actual vibration up through the spacecraft. So it was obvious the engines were running well. When the solids lit off, that was really quite a kick in the pants. I'm not really sure what I anticipated, but the only thing I've been able to equate it to, being a Navy pilot, is a catapult shot off an aircraft carrier. Most of the previous manned space launches have lumbered off the pad, but the shuttle doesn't do that. It gets up and moves.

Joseph P. Allen

Prior to his own flight on STS-5, the first fully operational shuttle flight, Allen served as entry CapCom, or spacecraft communicator, for Columbia's *reentry and landing.*

In the first days of spaceflight, when a spaceship would come back to Earth, it would fall into the atmosphere a bit like a rock. It would hit the top of the atmosphere so hard, causing the air around it to ionize; atoms would actually come off of the air molecules. When that happens, you can't send radio transmissions through this "sheet" ionization. It's a period known as a blackout. The spacecraft can't talk to the ground and the ground can't talk to the spacecraft. The ground also loses all of the data coming from the spacecraft, so nothing gets through.

In the early days this was anticipated, and the blackout period lasted for a fairly short period of time, only a couple of minutes. But the space shuttle is a flying machine and it doesn't come into the atmosphere like a rock. Nonetheless, it hits the top of the atmosphere at exactly the same speed, causing the same ionization effect. At Mach 23 (eighteen thousand mph) that's enormously fast, faster than a bullet. And because the orbiter is an airplane, it doesn't plunge through the atmosphere, it kind of glides through it for a longer period of time. So the ionization period lasts longer and blackout is on the order of five minutes.

And if a crew member happens to be talking at that time, their voice goes out. We know then that they're approaching and just call out, "Standby for blackout." During the wait of no communication you can almost hear people in Mission Control breathing. The next call that you hear in the mission control room is, "We have data, we have data." Suddenly the radios in the spacecraft start to transmit through the ionized layer.

It seemed to take forever in that first flight. We had never flown an airplane at that speed before, not even at half of that speed. We thought it would survive, though we didn't know. There was a huge sigh of relief when it did.

CHALLENGER

Technical problems and bad weather delayed the launch of the twenty-fifth shuttle mission six times before *Challenger* finally lifted off from the Kennedy Space Center's Pad 39B at 11:38 A.M. (EST) on January 28, 1986. It had been so cold the night before that icicles had formed on the shuttle's service tower and plumbing connections were frozen. The decision to launch was nevertheless made.

As *Challenger* rose from the pad and went into the familiar rollover, the intense flames in the right solid-rocket booster (SRB) began to eat through one of the O-rings that sealed two of its sections. Seventy-three seconds after liftoff, the seal ruptured, in effect turning the SRB into a blow-torch trained on the propellant-filled external tank, which then exploded.

Thousands of disbelieving spectators, including the parents of New Hampshire schoolteacher Sharon Christa McAuliffe, saw the spacecraft turn into a ghastly white flower as the external tank erupted, sending flaming and smoking wreckage across the azure sky.

The worst accident of the space age claimed the lives of astronauts Francis R. Scobee, Michael J. Smith, Judith A. Resnik, Ellison Onizuka, Ronald E. McNair, Gregory B. Jarvis, and McAuliffe. On February 3, President Ronald Reagan signed an executive order establishing a commission, which investigated the tragedy and made subsequent recommendations to the agency.

While the nation mourned and NASA and its contractors worked to correct a number of design and procedural flaws, the three remaining shuttles, and therefore almost all of the civilian space program, remained grounded for thirty-two long and gloomy months.

▼

Our nation is indeed fortunate that we can still draw on an immense reservoir of courage, character, and fortitude, that we are still blessed with heroes like those of the space shuttle *Challenger*. Man will continue his conquest of space. To reach out for new goals and ever-greater achievements, that is the way we shall commemorate our seven *Challenger* heroes.

Ronald Reagan, president of the United States

Flight Director Jay
Greene reacts to the
news coming in to
Mission Control that
Challenger has had a
"major malfunction."
It was the first time
American astronauts
had ever been lost
during a spaceflight.

CHALLENGER

Brian Perry, flight dynamics officer (FIDO)

The folks who reported to me who were doing the radar tracking saw a bunch of unexpected targets. All of a sudden there were a bunch of pieces where there wasn't supposed to be, so a lot of funny-looking data got splattered up on the plots. And that was my first indication that something was strange. . . . I remember Jay [Greene, flight director] making a call with some level of alarm in his voice, saying something like, "FIDO, trajectory?" At the same time I was getting a call from my navigator, saying they had all of this stuff going into the radar filter. So I was talking to them. If you listen to the tapes, it's a long time before I answered Jay.

At some point the telemetry dropped out, which wasn't unusual, but when it started to drag out for an unusually long time, Jay made an announcement about "Everybody watch your data carefully" or something like that. And that's where we started to get the feeling that something really wasn't right.

Richard L. McAfee, biology teacher

McAfee was one of the thousands of teachers across the country who had applied for NASA's "Teacher in Space" program.

I was teaching on the day of the disaster. I did not have a TV or radio in my room, so I was not aware of what had happened until some of my students told me. I hoped it was just a wild rumor, but I couldn't stop the tears. I'm sure it was the same for all of us who had applied as teachers. It was not just a national tragedy, it was extremely personal. A part of me, of all of us, was on that mission. It was a very personal loss. As I watched the McAuliffe family's reactions, I saw my own family in their places.

Craig Kennedy, pararescueman, 39th Aerospace Rescue and Recovery Wing

Kennedy was on helicopter assignment to ensure that the offshore waters of the launch area were free of boats, a standard procedure prior to any launch, and witnessed the tragedy from the alert point nearby.

We were watching it, and at first you could see something . . . you're in denial the whole time, but you could see something was definitely wrong initially, and you'd say, "Well, that could be just the solid-rocket booster separation," but then you saw those smoke trails. Everybody's standing out there—except the pilot who was in the helicopter because it was running—they just had their jaws wide open. Finally the pilot, who was in communication with Cape leader, started banging on the side of the helicopter to get our attention because it was just so loud out there. He said, "Come on, let's go." As we were getting in the helicopter I remember asking him, "Do we have survivors?" He answered, "We have a crash location."

It was a weird feeling. I'd say shock, grief, and adrenaline. I've never really experienced it before. Shock because what had happened shouldn't have happened. This was kind of a bulletproof process that we had come to really hang our hats on. And something like that happening was unthinkable. Then grief, because you knew what had happened, and more than likely there were seven people who were dead out there. And then adrenaline, because you knew if one of them was alive, you were going to have to go in and get them. As we were flying out, it seemed like we were moving in slow motion.

As I was putting on my wet suit in the back of the helicopter, I looked out the window and that's when I saw the first piece hit the water. It was about the size of a Volkswagen and it cleared the helicopter by about fifty feet, making a big splash in the water. While I was watching that, everybody else up front started noticing all the other stuff that was coming down. We had seen the impact point from about eight miles out, and we were going to make a couple of low passes to see if there were any survivors or any bodies. We decided quickly that if there were, then we, the pararescuemen, would have gone in the water and the helicopter would have gone north of Daytona. We would have taken our chances in the water. But as it was, we didn't see anything, so we made a couple of passes and then headed north while we waited for all the debris to fall.

When we got back after about half an hour, the water was just amazing because of all the hypergolic fuel; it looked like a giant witch's cauldron boiling. The water was boiling, then smoke was coming out of the bubbles. You knew that if anybody was in there, then they probably weren't going to make it, and you knew that if you were going to go in there, then you probably weren't going to make it either. . . . The helplessness of it all was what drove the way everybody felt. It was so catastrophic.

Challenger's solid-rocket boosters fly off in different directions, but no orbiter emerges from the cloud. Veteran shuttle watchers knew instantly what had happened.

Barbara Morgan, backup candidate for the NASA Teacher in Space Program

Morgan trained with Sharon Christa McAuliffe and the Challenger *crew from September 1985 to January 1986 at the Johnson Space Center in Houston.*

I sure feel lucky I got to know [the *Challenger* crew]. I learned so much from every one of them. I learned from them as teachers and as leaders. From Dick I learned that a great leader, a great teacher, leads more than he or she commands. Dick was the kind of leader where you always felt that you were right there with him and then when you finally got to where you were going, you realized that he was already there long before you.

From Mike I learned that a great teacher, a great leader, has great confidence in his or her student or team member. Mike took me flying in the T-38s [training jets] in the most exciting rides I'd had ever in my life. From Judy, I learned that you don't have to be all things at all times. Judy was phenomenal. She was the flight engineer, so she sat between Dick and Mike and monitored everything they did. Judy in work was very serious and very focused. It was all work. On the other hand, when it was not work time, she was a lot of fun to be around. From El I learned that a great teacher and a great leader lets his or her students learn alongside them. From Ron I learned great faith. He had great faith in a higher being, [in] himself, and in other people. He would often ask Christa and me about our students. He took a great interest in the children that we work with in our classrooms and would ask us a lot of questions. He had planned eventually to go back and be a college professor. From Greg I learned that a great teacher and a great leader makes the best of all situations. Greg had been bumped from several flights before he joined the *Challenger* crew. In fact, he joined the crew one month before launch, and they had been training together for over a year. From Christa I learned a lot of things, probably more than the other folks because I did spend more time with her. A great leader and a great teacher knows what's important and knows how to pay attention to what's important and let things go that aren't important. She was very, very good about balancing those things. What was really important to Christa were her students, their futures, their education, and their dignity.

Challenger's crew was as diverse as America itself. From left, Ellison Onizuka, teacher Christa McAuliffe, Michael Smith, commander Dick Scobee, Judy Resnik, Ron McNair, and Greg Jarvis, a civilian payload specialist who worked for satellite maker Hughes Aircraft.

Edward O'Connor Jr., colonel, 6555th Aerospace Test Group, Cape Canaveral Air Station

O'Connor directed the search, recovery, and reconstruction team supporting the presidential commission investigating the Challenger *accident.*

There was a tremendous shock as we all watched the accident occur. We were all looking at the sky where the vehicle was breaking apart and hoping we would see the orbiter come out from behind the cloud and do a return-to-launch-site. It was a difficult maneuver, but it was the one thing that you always held out hope for.

During the days that followed, time had no relevance. It was something that you were just totally immersed in. It took a dedicated chunk out of your life, out of your emotions, out of your heart, out of your very being. People made that sacrifice and worked to get us back in space again. When you saw the faces in the hallways, particularly immediately after the accident, everyone who had worked on that vehicle was wondering, "Was it something I did or something I didn't do? Something I overlooked?" Emotions were quite strong across the board, but everybody had to keep them bottled up until we finished the job.

DISCOVERY
RETURNS TO SPACE

With millions of Americans watching on television and the Cape Canaveral area crammed with a million more cheering it on, the orbiter *Discovery* carried the U.S. civilian space program back into orbit on September 29, 1988, after a thirty-two-month grounding in the wake of the *Challenger* disaster.

The launch delayed one hour and thirty-eight minutes due to lighter-than-expected upper atmospheric winds and the need to replace fuses in two of the crew members' flight-suit cooling systems. Suit repairs were successful and the wind conditions were deemed to be within a sufficient safety margin. Finally, the famous "Go" was given. The frequently interrupted countdown ended at 11:37 that morning. As *Discovery* rolled over and continued climbing, one number came to many minds: seventy-three. *Challenger* was lost seventy-three seconds into its flight, so that became the hurdle that had to be cleared.

To further make the connection, *Discovery* carried the same kind of satellite that its doomed predecessor had carried: a five-thousand-pound Tracking and Data Relay Satellite (TDRS) designed to serve as a communication link between shuttles and other spacecraft. Six hours after liftoff, and at an altitude of 184 miles, the TDRS was lifted out of the payload bay and fired into a 22,300-mile-high geostationary orbit, where it would remain over the same spot on Earth.

At the Johnson Space Center in Houston, about a hundred senior NASA officials and their guests held their applause until two minutes and twelve seconds into the flight, when the solid-rocket boosters fell away.

It was so important for the United States to reestablish its technical credentials in front of the world. It was a great team victory for everybody who worked on the flight.

Frederick "Rick" Hauck, commander

Back in space at last after standing down for nearly three years, the space shuttle is silhouetted against a bright Earth.

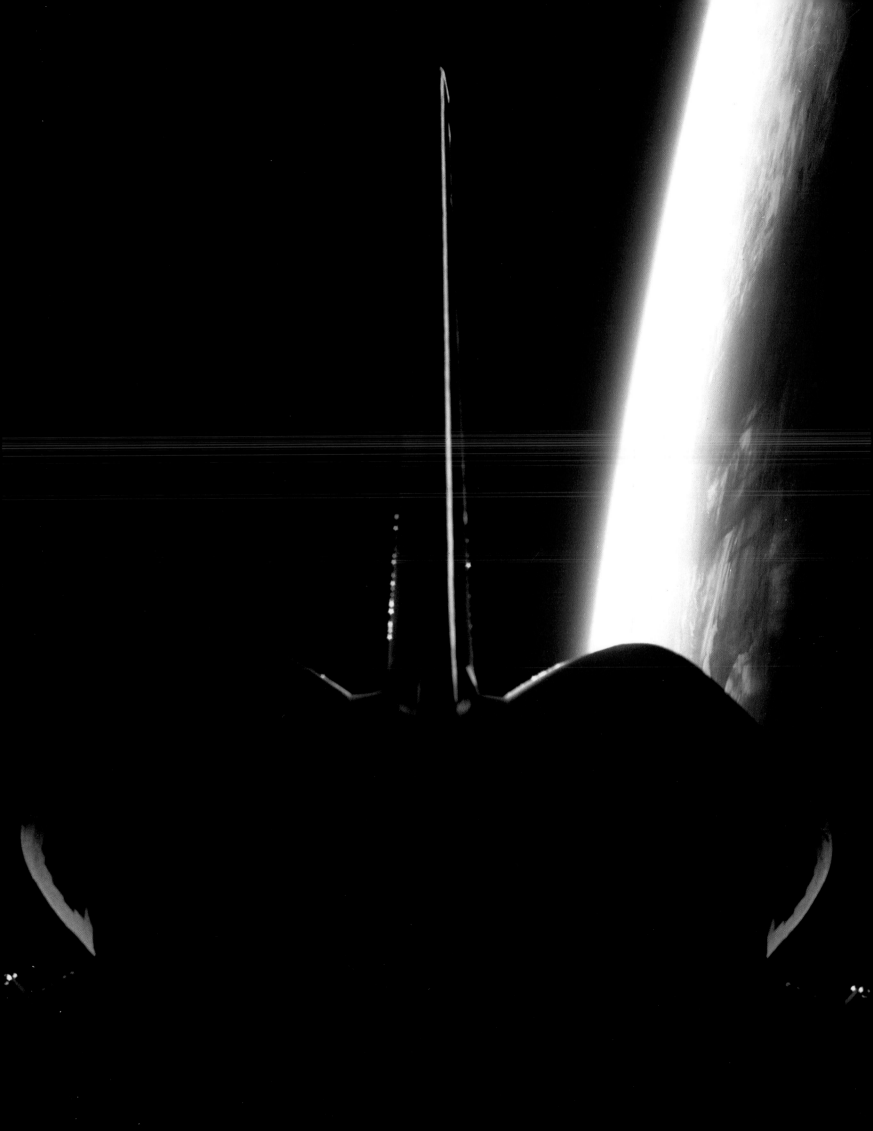

DISCOVERY

Richard H. Truly, associate administrator for spaceflight

In February 1986, then–Rear Admiral Truly, former shuttle astronaut, was appointed to lead the painstaking rebuilding of the space-shuttle program.

One of the decisions that I made was that when we named the crew for the first flight, whether this was going to take one year, eighteen months, whatever, we were going to name an experienced crew. We weren't going to fly any rookies and, furthermore, people were going to be in the seat in which they had previously flown. The commander of that mission would have to have been a previous commander. The pilot would be a previous pilot.

I took that same approach with a lot of other things. We decided that we were going to land during the day, not at night. We were going to land at Edwards Air Force Base, not at Kennedy Space Center. And so there was a whole series of ground rules and strategies that we laid out. The reason was twofold: One, it was to make our job easier and try to get as many of these things decided ahead of time, and two, it was to assure Congress and the country that we weren't going to fly until we were ready. But once we were ready to go flying, we would stack everything in our favor, which we did.

Frederick "Rick" Hauck, commander

I wouldn't describe flying this mission as heroic. I certainly felt that we were the focus of a lot of attention, and we got enormous amounts of mail before the flight. I remember getting a letter from a young man, I think he was in high school. I recall reading the letter and thinking, "I want to take this letter up to orbit with me." In it he said, "I'm fearful for our space program." I called his mother from our prelaunch quarantine and then spoke to him to ask his permission to read his letter during a call to Houston from orbit. I asked him what he meant by being afraid for our space program. In essence, he was saying that he hoped everything went well because if it didn't, he was afraid that we would give up. In some ways that epitomized what that flight was all about. If we had had a serious problem where people were injured or killed, I wonder what would have happened to our manned space program.

Below

Nearly a million people are on hand in Florida to cheer *Discovery* into orbit. A traditionally popular viewing site, the NASA causeway over the Banana River supports throngs of viewers.

Below right

Discovery waits on the pad. After *Challenger*, NASA made changes to the shuttle that rendered the vehicle safer and more capable.

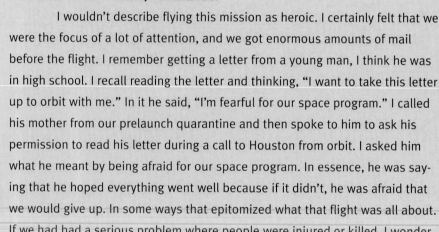

Robert L. Crippen, astronaut, deputy shuttle manager

Getting the shuttle back flying safely again was high on my priority list, along with all the other NASA management. People were so nervous after that accident, I knew it was not going to be easy because they would be coming out of the woodwork looking for reasons not to go fly. . . . I was very confident in the vehicle and in the team. But it's much more difficult to sit in the management chair at liftoff than to sit in the cockpit because you know you can't do anything there. At least in the cockpit you feel that you might be able to do something. . . . I know that the *Challenger* crew would not have wanted that to be the end of the shuttle program. So getting it back flying safely was extremely important to me. Truthfully, as I look back over my career, that's probably more of a highlight in my mind than flying STS-1.

Commander Rick Hauck (left) and Pilot Dick Covey
review checklists on *Discovery*'s flight deck. For
the return to orbit, NASA chose its first all-veteran
shuttle crew.

Richard H. Truly

It was like climbing Mount Everest fourteen consecutive times, it seemed
to me, to get to launch day. And it was a beautiful day. I mean, my God, you could
see forever. And dadgum if we didn't end up with some upper wind condition, and
I thought we were going to have to cancel. And Rick [Hauck] and the crew were
out there on the pad, and the launch window was closing over in Africa. We had
the whole nation watching. If we would have had to abort that day, we would
have looked like we were chicken to go fly, and I didn't want to go do that. But I
would not change the emphasis on safety that I had tried to put back into the program. . . . Very near the close of the launch window, suddenly the winds went our
way. We started the countdown and the guys launched. I just held my breath for
eight and a half minutes until the main engines shut off. It was exhilarating. It
was an unbelievable feeling.

Frederick "Rick" Hauck

When we got into the final count and the main engines lit and then the
solid rockets ignited, I was totally focused on business and looking at the gauges
and flight parameters. Within the first twenty seconds after liftoff we did have an
alert message associated with the fuel cells. It turned out that it was not a serious problem—it was not a life-threatening problem from the beginning—but here
we are, just lifting off and we have this light illuminate and a message on the
computer saying, in essence, there's a problem. A chill went down my spine. But
it was back to business and up we went. . . .

Once the main engines cut off, we were off the tank; we did our OMS
[orbital maneuvering system] burn and achieved orbit. To me, at that point we are
in a very benign environment. Most of the bad things that could happen to you
would have happened by now. . . . It's a rather serene experience, this noiseless,
silky smooth, gliding over the earth, circling the earth every hour and a half. This
wasn't the time to reflect on danger, rather it was a time to savor the beauty of
what we saw looking down at the earth, and to revel in the magic of being free of
gravity's grip, and enjoy the camaraderie of your friends—listening to classical
music while passing over the Himalayas. It's an experience I'll cherish forever.

Jane Ira Bloom, musician and composer

*Bloom translated her experience of viewing the STS-26
launch and landing into a suite of musical compositions for
the NASA Art Program in celebration of the return to flight.*

We all dream about the possibility of flying in
space, but as artists we live in a world of imagination. That's
a gratifying place to live in, to breathe in. To imagine what it
might be like is also a wonderful experience for an artist. . . .
There were some very exciting concepts that I tried in the
piece that I wrote for NASA that I had done for the first time
and later developed.

One of them was taking some elements of a signa-
ture movement that I use in my own playing and orchestrat-
ing it for whole sections of brass instruments where they
actually move in unison to create Doppler-like effects. The
other aspect was having instruments in the orchestra placed
in an omnidirectional way around the hall, so the sound
came at you from many different angles. . . . I got the most
wonderful comments from a few people who were in the
audience who were physicists and engineers. They were fascinated by this idea because they knew where it came from.
Astronauts, too, were thrilled technically by what was going
on. They "got" the idea of what was resonating between
things understood scientifically and things expressed artisti-
cally. . . . I've carried that experience in 1988 through my
music and my work ever since. It was pivotal for me as an
artist and as an observer of American history.

▼

> After the accident, I
> always felt my job was
> to look forward. And it
> was the hardest thing
> I've ever done. It was
> ten times harder than
> flying the space shuttle.
>
> **Richard H. Truly**

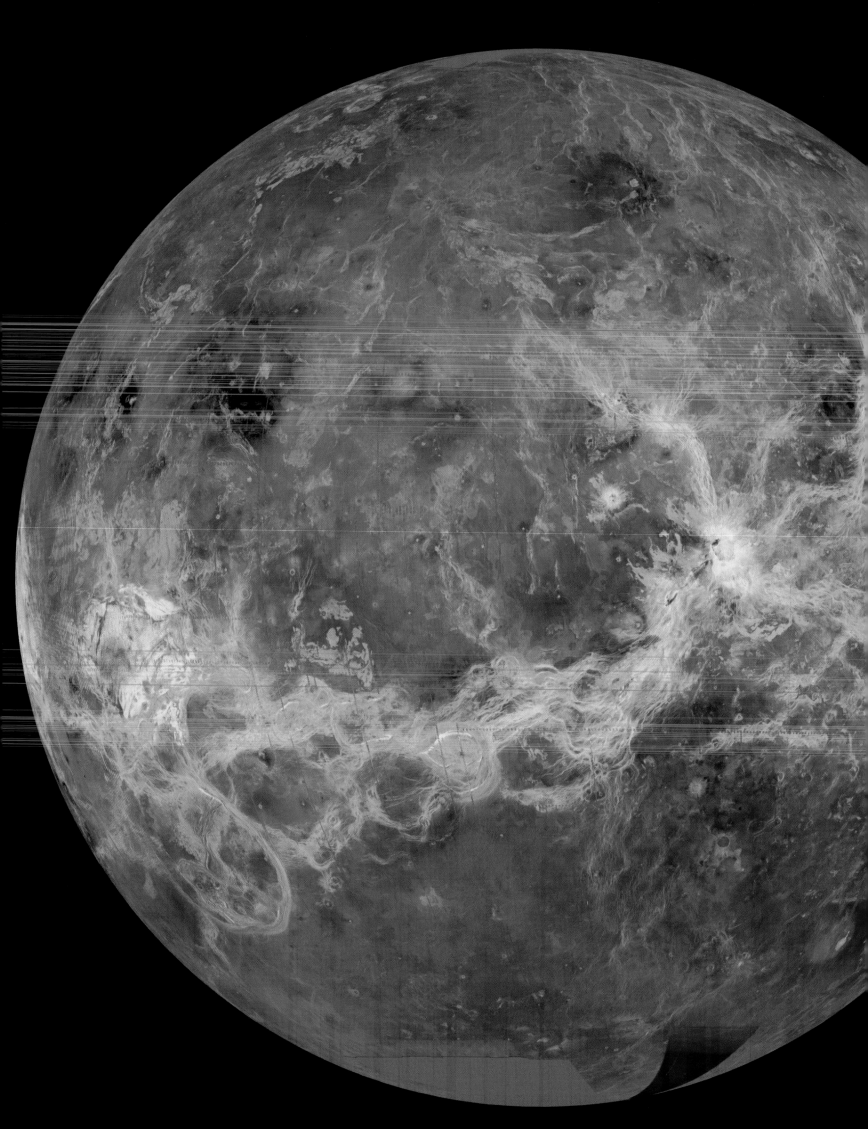

EXPLORING
DISTANT WORLDS

The urge to explore the unknown, from inside atoms to the farthest reaches of the universe, is a uniquely human instinct. It has sent brave men across uncharted seas from the time of the ancient Greeks and Vikings to the Golden Age of Exploration during the fifteenth through seventeenth centuries and, in the twentieth century, to the icy poles that crown the opposite ends of this planet.

Aristotle mused about the infinitely small and infinitely large, but it was not until the end of the sixteenth century and beginning of the seventeenth that the invention of the microscope and telescope vastly extended the vision of the infinity seekers. For those investigating the heavens, Galileo's telescopes, which were roughly four feet long, made of wood, and held lenses barely an inch in diameter, bridged the ancient gap between the human eye and what lay beyond its natural limitations. Since then, astronomical study has very largely been about extending that range, most recently with a powerful new partner: the rocket. The rocketeers not only have sent extraordinary observatories such as the Hubble Space Telescope (HST) and Cosmic Background Explorer (COBE) and hundreds of other science missions above the atmosphere, they have sent far-ranging robotic explorers to reveal a richly detailed portrait of most of the solar system. The results have been so spectacular that they overwhelm superlatives.

Fueled by the Cold War, solar-system exploration started at the Moon and then continued farther to Venus and Mars. Through it all, scientists applied an underlying logic to exploration: Scout successive planets, beginning with a craft flyby that streaked

The face of Venus as revealed by Magellan's cloud-penetrating radar in 1991. The surface is invisible to conventional cameras.

Jupiter's moon Io

orbits close to the giant planet in this Voyager 1 image from March 1979. Features on Io as small as 38 miles across appear in this view, which was taken from a distance of 5 million miles.

▼

Mysterious Venus's veil

was finally penetrated, and what turned up was a starkly beautiful land-scape with features found nowhere else in the solar system.

past a target to gather information; circle the planet, which would return more data; finally, make a landing that could afford the most detailed, up-close investigation.

Mariner 2 was the first visitor from Earth to successfully reconnoiter another planet when it sped past Venus on December 14, 1962. During its forty-two-minute "encounter," it measured the cloud-shrouded, carbon-dioxide-soaked planet's temperature (roughly eight hundred degrees Fahrenheit, or twice as hot as expected).

Far more important, Mariner 2 proved that a spacecraft could be sent to another planet on an accurate trajectory provided it worked properly and the position of the destination planet at any given time, called its ephemeris, had been meticulously calculated. Determining the ephemerides of planets, moons, asteroids, and other inhabitants of this solar system entails calculating each moving body's speed, gravitational influence, and other characteristics to figure out precisely where it will be at any given instant. This astronomical art form is underappreciated by laypersons, but it is cherished by engineers and space scientists who know that they can't get to a target unless someone skilled in this area tells them precisely where it will be when their spacecraft enters its neighborhood.

Eight Mariners, most of them successful, followed Mariner 2 on visits to Venus, Mars, and Mercury through 1975. Meanwhile, NASA was looking past Earth's closest neighbors to the giant gaseous (actually liquid) planets Jupiter and Saturn. On March 2, 1972, a five-hundred-seventy-pound probe named Pioneer 10, developed by the Ames Research Laboratory, left Cape Canaveral heading for Jupiter. Thirteen months later its twin, Pioneer 11, left for Saturn. Both made it through the asteroid belt and survived Jupiter's deadly radiation to send back a stream of spectacular pictures, including the first close-ups of Saturn's rings, and data on their atmospheres, magnetic structures, moons, and interiors. By September 1995, just prior to mission operations terminating, Pioneer 11 was four billion miles from Earth and heading into interstellar space.

The Pioneers were pathfinders for the Jet Propulsion Laboratory's follow-on missions: Voyagers 1 and 2. The first barreled past Jupiter in March 1979 and then went on to Saturn for a close encounter in November 1980. It sent back a trove of data on both planets' composition, atmosphere, moons, and other phenomena before streaking out of the system.

Its sister, Voyager 2, lifted off from Cape Canaveral on August 20, 1977, and spent the next twelve years conducting an unprecedented four-planet "Grand Tour" of Jupiter, Saturn, Uranus, and Neptune that amounted to the most outstanding feat of exploration in history. For the sheer magnitude of data from all four planets and their retinues of moons—several of them first spotted by the Voyagers, and others as complex and varied as worlds themselves—the Grand Tour was unsurpassed.

In May 1989, the year Voyager shot past Neptune, follow-on missions were sent to Venus and Jupiter using data supplied by the Mariners and Voyagers. The Venus encounter was completed by an orbiter named Magellan, whose radar was able to penetrate the cloud cover for the first time and return thousands of high-quality images of virtually the whole planet. Mysterious Venus's veil was finally penetrated, and what turned up was a starkly beautiful landscape with features found nowhere else in the solar system. Galileo, a heavy orbiter carrying a probe, was launched by a shuttle on October 18, 1989, for a return to Jupiter. It not only circled the planet repeatedly, sinking the probe into its atmosphere like a harpoon into a whale, but scrutinized several Jovian moons. One of them, Europa, seemed to have a frozen-over sea of water ice, meaning possible life. Galileo was still working as the twentieth century turned into the twenty-first, even as another heavy explorer named Cassini was bearing down on Saturn to start the same kind of operation.

Closer to home, Hubble, launched by the shuttle *Discovery* in April 1990, was astounding its users with startlingly clear images of exploding supernovas, gigantic star incubators, "close-ups" of galaxies being born and colliding, billions of other galaxies in clusters so thick they look like stars, and much more.

Together, the new explorers have not only revolutionized humanity's view of the universe, but they have helped it to set its own migratory course. And they have shown what humankind is capable of when it sets its sights on the zenith.

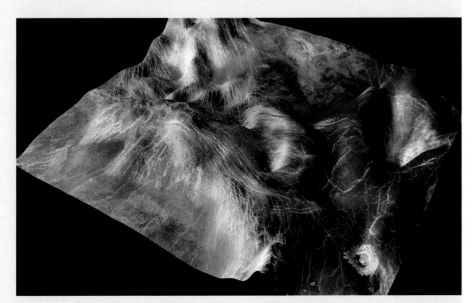

This computer-generated view of the Venusian lowland plains in Sedna Planitia shows "coronae," unique circular depressions and their associated fracture patterns, which tend to occur in linear clusters across the planet's major tectonic belts.

PIONEERS 10 & 11

B y the end of the 1960s, much remained unknown about interplanetary space beyond Mars and the distant planets of the outer solar system. In 1969, the National Academy of Sciences recommended exploring Jupiter, Saturn, Uranus, Neptune, and Pluto in what would eventually be called a Grand Tour.

NASA's Ames Research Center and a company called TRW developed two spin-stabilized explorers—craft that maintained their balance by spinning like tops—to size up Jupiter and Saturn. Like Ames's previous explorers, they would be named Pioneer—in this case, Pioneer 10 and its twin backup, Pioneer 11. If these two cosmic canaries made it through the asteroid belt without taking a devastating hit, and survived Jupiter's murderous radiation, other explorers would follow them past the two giants to Uranus, Neptune, and beyond.

Pioneer 10, sporting an aluminum plaque with a message of its origin, left for Jupiter on March 2, 1972, and in December 1973 became the first spacecraft to directly observe the planet. It returned close-up pictures, plus an enormous amount of data on the Jovian moons, radiation belts, and magnetic field. It also discovered that Jupiter is predominantly a liquid planet.

Pioneer 11 left Earth on April 5, 1973, flew past Jupiter in December 1974, and in 1979 completed the first direct encounter with Saturn. It discovered two small moons and a new ring, reported Saturn's strong magnetic field and magnetosphere, and revealed its moon Titan, which is larger than Mercury, to be too cold to support life.

▼

We never dreamed that Pioneer would survive this long in space. It was our fondest hope that it would last at least five years to get us by Jupiter and make some measurements beyond Jupiter. Those fond hopes have carried on over into twenty-five years. This is truly one of the major scientific and engineering achievements of the twentieth century.

Richard O. Fimmel, project science chief

Engineers at TRW put the finishing touches on the Pioneer 10 spacecraft before its March 1972 launch. The twin Pioneers were true to their names: They were the first spacecraft to cross the asteroid belt and the first to explore Jupiter and its large moons. Pioneer 11 went on to Saturn, passing within 13,000 miles of the ringed planet in 1979.

PIONEERS 10 & 11

Devrie Intriligator, investigator, plasma analyzer

In addition to light, the Sun emits a stream of charged particles known as the solar wind, which interacts with each planet's magnetic field.

The existence of the solar wind was inferred, in part, because comet tails always point away from the Sun. Earth, Jupiter, and the other planets are obstacles in the solar wind's flow, just as a comet is. Since the solar wind is supersonic, flowing at speeds of nearly one million miles per hour, a bow shock forms upstream of these obstacles and a tail forms downstream, much like waves around a rock in a fast-moving river.

One moment that stands out in the Pioneer 10 mission was when our plasma analyzer detected the first evidence of Jupiter's bow shock. It detected the bow shock on a number of occasions. We did not expect Jupiter's bow shock and space environment, or magnetosphere, to be so responsive to changes in the solar wind. When the solar wind blew fast and strong, it forced Jupiter's bow shock much closer to the planet and compressed its magnetosphere. As a physicist, it was a real thrill to spot that kind of phenomenon, since it illustrated that we understood the physical mechanisms at work.

Richard O. Fimmel, project science chief

Since Pioneer 10 would be the first spacecraft to fly through the asteroid belt, scientists had to take into consideration the special perils of the trip.

Our concern was that asteroid belt might almost be like a whirling sea of gravel. How would a spacecraft get through that? It could be pulverized. In order to avoid the asteroid belt, you would have to climb up over it. By going through the belt, however, we would not need a giant rocket, which at that time only the Russians had.

There were known asteroids, large ones a mile wide or longer, that had been tracked by telescope from Earth. Their orbits were known. We had our ephemerides specialists chart the trajectory of the spacecraft and we planned the launch accordingly to avoid at least the known hazards. Two detectors were onboard Pioneer. One was an asteroid meteoroid detector (AMD) that could optically detect any large objects that would pass close by. But fortunately, the whole path was charted so well that the AMD instrument didn't have anything to observe. Another instrument was a micrometer detector that could detect smaller particles that might impinge upon the spacecraft and erode it, much like the process of sandblasting. But as it turned out, we detected more micrometeoroids before and after the asteroid belt encounter than during the asteroid belt encounter.

To explain why, think about gravity, which is the force of attraction depending on the magnitude of mass of the body. The large asteroids, circling for untold millions of years around Earth, have bumped into the cosmic dust particles and with their gravitational force of attraction have served as so-called "cosmic vacuum cleaners," sweeping the area clean. In retrospect, it was such a simple concept that you'd say to yourself, "Why didn't we think that would be the case?"

Having Pioneer travel through the asteroid belt was a major achievement for our space effort because all subsequent missions could follow in Pioneer's footsteps and avoid the expense and waiting for the larger rockets needed to climb up and over the belt.

As Pioneer 10 disappeared behind Jupiter, it was not certain whether the spacecraft's electronics and instruments would survive the bombardment of intense radiation.

When the spacecraft came out from behind Jupiter and the first signal arrived, I saw that several of the instruments had taken hits and were switched into the wrong mode. Because we were moving away from Jupiter at a high velocity, we had to get them switched back or we would lose the close-up data from Jupiter. I recall charging down the hall and pushing people out of the way to get to the control room in order to send the commands out to reconfigure the spacecraft instruments. I managed to get the commands sent in less than five minutes.

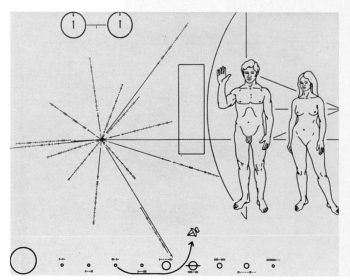

A plaque fixed to the first interstellar spacecraft shows its human builders and where they live. Pioneer 10 will reach the stars of the constellation Taurus in about two million years.

Jupiter's Great Red Spot
and the shadow of its
innermost moon, Io,
show clearly in this
image taken by Pioneer
10 in December 1973.

Richard O. Fimmel

Pioneer 10 was in Jupiter's radiation belt for a fairly long time, and there was some radiation damage. We decided that if we targeted Pioneer 11 farther out, we wouldn't see as much or get the measurements we wanted. But what if we targeted closer in? We would get more intense radiation, but radiation is a matter of dosage. By coming in closer and at a much higher velocity—now that we knew how intense the radiation was at Jupiter—we could calculate exposure in advance. The dosage would be less. And it turned out to be true. Pioneer 11 went through without any radiation damage.

We also decided that if we made another equatorial pass, we would not see the poles of Jupiter, so we retargeted Pioneer 11, in flight, to fly under Jupiter's south pole and up over its north pole, then loop back past Earth and go on out to Saturn. . . . We knew that the Voyager team wanted to be the first to get to Saturn, but in retrospect that was really not important because they were able to gain enormous benefits from the Pioneer 11 Saturn flyby.

Tom Gehrels, imaging team

The pictures of Jupiter were so exciting, and the results from the particles and fields experiment were so fabulous, that NASA was looking for people to tell the world. I was asked to go spread the news, and I literally went around the world. Invitations came from unexpected places, like from Prince Charles at Buckingham Palace and Indira Gandhi, the prime minister of India. We went to the crown prince of Japan and his son, and they were fascinated. Then we met with the king of Sweden and the Dutch royal family. We went to scientific academies and big meetings, too, such as to see twenty thousand Japanese children in the stadium of Hiroshima. There was such great interest because the Pioneers had made the very first passage by Jupiter and Saturn. It was the first time we'd been to the outer part of the solar system.

Tom Gehrels

The instruments on Pioneer 10 had been designed on a model of Jupiter's radiation obtained largely by ground-based radio astronomers. But this radiation model was changed while we were already flying to Jupiter. It was calculated to be much more serious, much more energetic, than what we had designed the spacecraft to handle. So it gave us great concern and led to predictions that were rather dim for the day of Pioneer 10's encounter with Jupiter, set for December 3, 1973 at about 6:30 P.M. (PST). We anticipated that our instrument, the imaging photopolarimeter (IPP), the most radiation-sensitive instrument onboard would start doing weird things at about noon of that day. And indeed, at noon, the instrument became very erratic, and began searching in areas of the sky where it wasn't supposed to be searching.

The next prediction was that at about 3:00 P.M. nearly all of the instruments, including the sensor for orienting the spacecraft, would be damaged by Jupiter's excessive radiation. And later in the afternoon, around 3:30 P.M. or so, all of the spacecraft communications would go erratic and possibly be destroyed.

And, indeed, erratic signals were received at these times. We were quite concerned. As Pioneer 10's trajectory took it behind the planet Jupiter, all signals vanished. We knew this would happen because the radio signal was occulted while the spacecraft was behind the planet. The big question was how, or if, the radio signals would reappear after Pioneer 10's exposure to this unexpected high radiation.

But as the spacecraft emerged from behind the planet and the radio signals once again reached Earth, we determined that all of the instruments escaped major damage. This was the most astounding thing, especially for our IPP.

So the question started floating among all of us investigators: Why had the instrumentation survived? We then learned something that we weren't aware of. The engineers had built in a much bigger margin error than we scientists would have in the amount of radiation that the spacecraft could handle. And that saved the day.

VOYAGER'S GRAND TOUR

The greatest voyage of space exploration—the Grand Tour—had its roots in two otherwise unrelated discoveries: that Jupiter, Saturn, Uranus, and Neptune would be in rare alignment during the 1970s and 1980s, and that a planet's gravitational pull can increase the speed of a spacecraft.

These two facts combined meant that a spacecraft from Earth could encounter the outer planets in succession. And by using "gravity assist," beginning with massive Jupiter, it could economically "slingshot" its way from one planet to the next to reach its last stop, Neptune, in twelve years instead of thirty.

Following the success of Pioneers 10 and 11, two Jet Propulsion Laboratory spacecraft, Voyager 1 and Voyager 2, left for planetary flybys in the summer of 1977, with eleven science experiments onboard that would be run by ninety-five scientists. What the craft sent home was stunning.

Voyager 1 streaked past Jupiter in March 1979 and Saturn in November 1980, sending back the first detailed pictures of the giant planets. It then headed out of the solar system and, in February 1991, snapped a "family portrait" of several planets.

Voyager 2, partly deaf and arthritic, returned more data from Jupiter and Saturn, and then became the first visitor from Earth to encounter Uranus in January 1986 and Neptune in August 1989. In both cases, it whispered on low power an encyclopedic amount of data to elated scientists, including the discovery of a magnetic field, nine rings, and ten new moons at Uranus and six new moons and rings around Neptune.

▼

I think all of us who were on Voyager feel that it's hard to imagine any other mission with the wealth of discovery and diversity that this one mission provided.

Edward C. Stone, project scientist

Saturn's rings show their dazzling complexity in this close-up photo snapped by Voyager 2 in 1981. Colors have been exaggerated to bring out subtle differences in the chemical makeup of the icy ring particles.

VOYAGER

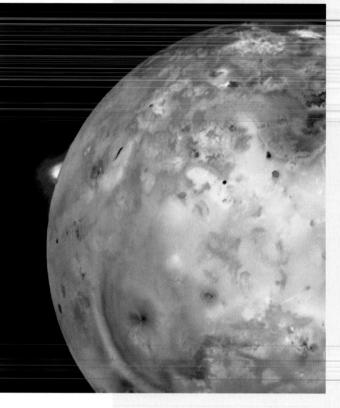

Voyager catches a volcano in the act of exploding on Jupiter's sulfurous moon Io. Active volcanism on a world so different from Earth was one of Voyager's great surprises.

Opposite

The atmosphere of Uranus glows in false color in this Voyager 2 image. By superimposing the planet's latitude-longitude grid, scientists could see that the atmosphere circulates in the same direction as the planet rotates. The planet's south pole was pointing toward Earth at the time of the flyby in January 1986.

Opposite, far right

A backward glance view of Saturn from Voyager 1 in November 1980. Voyager 2 went on to explore Uranus and Neptune.

Edward C. Stone, project scientist

One moment that certainly stands out was the discovery of the volcanoes on Io. That really set the tone for the rest of the mission—expect the unexpected. It broadened our expectations dramatically, that nature was really providing a whole set of diverse bodies out there waiting for us, and that no matter how hard we thought about it, I think we were always surprised.

One particular image of Io was originally taken for navigation purposes, which means a picture of the satellite is taken against the star background and is then overexposed so you can see the stars. That process made these little faint plumes just jump right out. And that was the place where the volcano was discovered.

They called me down to look at this thing because here was this big object on the limb of Io and it was quite spectacular. The first thing they checked was whether it could be another satellite behind Io. They did additional checks, but it wasn't long before we realized this was probably a volcano.

Another key piece of information became obvious at the same time. The infrared instrument that had been measuring the temperature of Io as a function of wavelength had noticed something peculiar. If you have a normal type of object, you expect the temperature of the surface to be the same no matter what wavelength you measure in the infrared. Well, here it wasn't. And basically, Rudolph Hanel [principal investigator, IRIS, Infrared Interferometer Spectrometer and Radiometer] said there were three possibilities: One, that there was some kind of mineral on the surface that had very peculiar absorption properties unlike anything anybody had ever seen before; two, there might be an instrument calibration problem, but everything else Hanel had looked at made sense; or three, that there was more than one temperature on the surface of Io. We just sat there totally puzzled. Of course, number three was the right answer. Hanel had been looking at a hot spot, the very hot radiating surface of a lava lake. It was such a stretch of our experience to realize that this little moon was the most active object in the solar system. It was just hard. It was such a change in paradigm.

Stamatios Krimigis, project scientist

Krimigis served as principal investigator for the Low-Energy Charged Particle (LECP) detectors that measured particles trapped in the radiation belts of the outer planets.

By having an instrument on Voyager, I felt that I was actually sitting on the spacecraft. I had these "eyes" that would actually see every proton and sulfur ion that was hitting the detector. Of course, that wasn't literally true, but it was how I felt. For me it wasn't just seeing data. I felt the radiation fields and imagined them as visual objects in my head. I would sit there and look at the data and say, "Ha! It looks like we ought to move our detector a little faster." And I would write up a command and stick it in the system. Then forty minutes later, I would watch the data to see what changed, and how it changed. The planetary encounters were sort of a continuous high for several weeks at a time.

Douglas G. Griffith, Neptune encounter preparations manager

Having held various mission operations manager positions between 1974 and 1989, Griffith finds it difficult to singularly describe his most exciting moments on the Voyager mission.

You almost have to break it down to each planet. For Jupiter it would have had to have been the discovery of volcanoes on Io. On Saturn, it was the complexity of the rings. They were much more complex than anybody ever imagined. At Uranus, the really interesting aspect was not only that the planet tipped on its side, but that it had this highly movable magnetic field that bounced back and forth, making it very unique. Another aspect about Uranus was the satellite called Miranda, which was spectacularly different than anything ever seen before. It has such extreme terrain on it, it looks like it was just blown apart then pulled back together by gravity. At Neptune, probably one of the most interesting things was finding out that the big satellite called Triton had geysers—liquid-nitrogen geysers that come up from its surface.

Carolyn Porco, ring specialist

We don't know how most of the stuff in the rings came about. That's pretty humbling. We're not sure how old the rings are and how long they'll stick around. There's evidence that suggests they're about a hundred million years old, but we're not sure if that's correct because we don't know the physical characteristics of the particles that make up the rings. They range in size from fine powder you might ski on in Utah to ice-bodies the size of houses.

We need to understand the rings because the solar nebula, from which the planets formed, was at one time a disc. We want to know how a disc works so we know how the planets formed. The rings are a local example of a celestial disc. It's not a perfect model, but it's the best we have.

Douglas G. Griffith

Voyager had a suite of eleven scientific investigations. In addition to photographic instruments, there were infrared instruments, atmospheric instruments, particle instruments, and instruments that heard radio emissions—all things you could not see visually. When Voyager went close to the planet and hit a couple of small particles that came up from the various rings, you could hear them pepper the spacecraft because an instrument had recorded it. Sometimes people get hung up and only want to see visible types of evidence. But the things you can't see visually can tell you an awful lot more about the whole makeup of the planet.

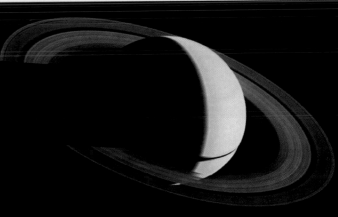

Carolyn Porco

I was sitting in the room at the Jet Propulsion Lab that had just been the scene of bedlam when Voyager was making its way through the Saturn system. People were there at all hours of the day and night. There were candy bar wrappers and coffee cups strewn all over and ashtrays overflowing with cigarette butts. We were all wracking our brains trying to figure out what the hell these pictures all meant. Now the encounter with Saturn was over and everyone had gone home. I was sitting there one night all alone watching the monitor that had displayed all these outrageous pictures. But now the monitor showed Saturn getting smaller as Voyager headed farther out. The Sun cast this enormous dark shadow over the far side of the planet and the rings. I felt as though I was on that spacecraft, gliding over this supernatural-looking system.

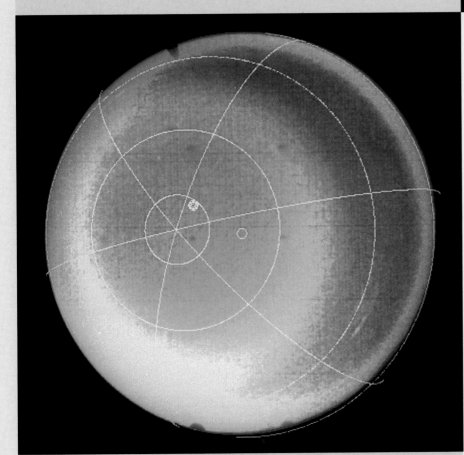

VOYAGER

Edward C. Stone, project scientist

When the Voyager mission was first being designed in the early 1970s, the moons of the outer solar system were not regarded as being nearly as interesting as they turned out to be. The expectation was that these small worlds would be frozen relics left over from four and a half billion years ago. It was not at all obvious that there would be such diversity. Now, of course, we're much smarter, and we realize that every one of them is different.

Saturn's tiny moon

Hyperion (three views shown here) may have been knocked out of its original orbit around the planet by a meteorite impact, which also pocked its surface. Voyager 2 took the farthest image from a distance of 740,000 miles and the closest from 310,000 miles away.

Stamatios Krimigis, project scientist

Fred Scarf, who had the plasma wave instrument on Voyager, sometimes took a dim view of all the Voyager images and the attention they got. He'd say, "That's fairly simple stuff. The electromagnetic radiation is where the real action is." And we would sit around and say, "Yeah, it's too bad we don't make pictures." So Fred said, "There's got to be a way to pick up some of these plasma waves and get them into the audio range." So he and his co-investigator thought they would feed the signal into an Apple music synthesizer. Those were the days when Apple was just coming out with personal computers, and there was one at the Jet Propulsion Laboratory. Some of the engineers said, "Why don't we feed these signals in there and see what comes out the other end?" And they got these very unusual sounds coming from the synthesizer being driven by the plasma wave data from Jupiter. You could hear the changing pitch, and this whistling sound. It was a big hit with the press.

Ann Druyan, creative director of the Voyager interstellar message

Voyager 1 and Voyager 2 each carry a twelve-inch, gold-plated copper phonograph disk containing a diverse collection of sounds and images representing life and culture on Earth. Druyan was the creative director of the project group, chaired by her future husband, astronomer Carl Sagan, and was involved in selecting from various sources the sounds of the planet: music, nature, civilization, language, and more.

When I would call people up—whether I was looking for the sounds of a Bronx schoolyard, or whale greetings, or an F-111 flyby—and told them what I was doing and why I needed the sounds, they would be incredulous. You can't blame them for being skeptical. I developed a little forty-second rap that I would try to get out as soon as possible before people either hung up or asked, "Are you nuts?"

I would just tell them, "My name is Annie Druyan and I am working with Carl Sagan and the National Aeronautics and Space Administration on a project to send sounds, greetings, music, and images of the planet Earth on the two Voyager spacecraft that will be launched in August and September of 1977. We don't have very much money but what we want to do is to put your cricket song or your playground sounds on this record. And if you help us do this, you will be touching something that conceivably might touch something else a billion years from now. It is our hope that the record will one day be intercepted by spacefaring extraterrestrials. But even if that doesn't happen, it is a message to the people of the Earth about how much we all have in common and how much we have been able to accomplish as a species."

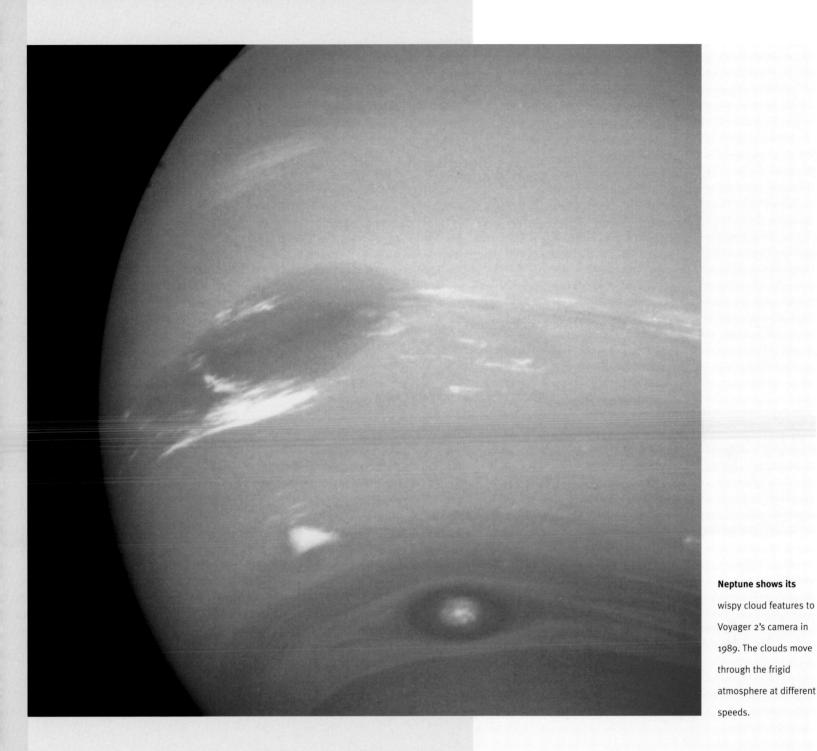

Neptune shows its wispy cloud features to Voyager 2's camera in 1989. The clouds move through the frigid atmosphere at different speeds.

Ann Druyan

If the Voyager engineers are correct in their assessment of the interstellar message's impossibly long future, then that means that it will outlast everything we know. A billion years from now there will be no pyramids, no Great Wall of China. The actual shape of continents will be so radically different, through just the natural, geological processes, that if we were to approach the earth a billion years from now, we wouldn't recognize our own planet from space. Of course, Voyager was launched in 1977, when the nuclear arms race was going at full tilt. The prospects then for human future were, I think, somewhat more dim, more pessimistic than they are now. . . . The idea of having a cultural artifact of who we were and of our most benign and unselfish acts of creativity, that could leave the earth, was just the most inspiring thing.

HUBBLE: CYCLOPS IN THE SKY

Astronomers have long known that studying the sky from under the atmosphere is like studying a diving board from the bottom of a swimming pool. Both efforts suffer from distortion. That is why earthbound telescopes are built on high mountains and why astronomers have wanted to place a telescope above the atmosphere ever since doing so became possible.

Scientists got their wish in April 1990, when the shuttle *Discovery* carried the Hubble Space Telescope (HST)—named for the astronomer Edwin P. Hubble—to a 330-mile-high orbit. Despite a series of problems—the most notorious being blurred vision caused by an improperly ground mirror that had to be fixed by a repair mission, *Endeavour*, STS-61, almost three years later—the HST has profoundly affected humankind's view of the universe. It doesn't see farther than other telescopes, but it does see far more clearly.

This clarity has provided the world with millions of new and insightful images of stars in various stages of life and death and billions of galaxies, some waltzing into slow-motion collisions of unimaginable force. Hubble has given us a phantasmagoric, Disneyesque image of star incubators—immense pillars of hydrogen and dust—in the Eagle Nebula that were creating stars under colossal pressure, and an eerie image of Saturn's great white spot that could have been painted by the artist Seurat. It took unprecedented, richly detailed close-ups of Supernova 1987a in its death throes. It is a time machine that has discovered black holes and sees almost to the edge of the universe.

▼

Hubble touches people because it symbolizes the power to look far enough to find out what life means here on Earth.

Story Musgrave, payload commander, STS-61

Astronaut Kathryn Thornton, secured on the end of the space shuttle's robot arm, makes a service call to the Hubble Space Telescope during the STS-61 mission in 1993.

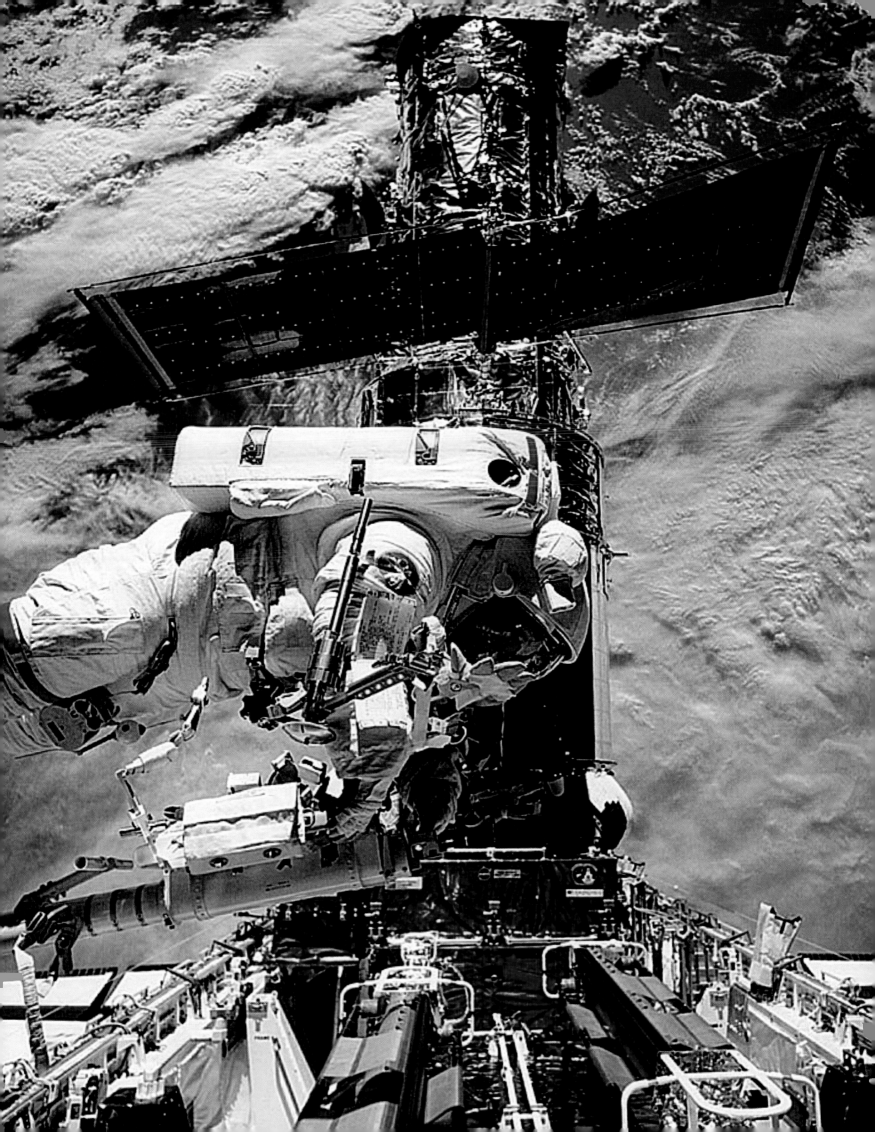

HUBBLE

Astronauts Kathy Thornton and Tom Akers practice installing the Wide Field Planetary Camera into a mock-up of the HST at the Neutral Buoyancy Simulator.

Kathryn C. Thornton, mission specialist, STS-61

As part of a seven-person crew, Thornton helped repair and service the Hubble Space Telescope during STS-61 activities, which included installation of the Corrective Optics Space Telescope Axial Replacement (COSTAR) to adjust for the aberration in the HST's primary mirror.

STS-61 was the first of a series of preplanned service missions for the Hubble Telescope, intended to be conducted every two or three years in order to upgrade its scientific instruments and replace any improperly working components—generally, to treat Hubble like a laboratory, where you would go in, make changes, and fix things from time to time. When Hubble was launched, and it was discovered that the mirror was not quite right, this planned service mission included a few unplanned tasks, including installing corrective optics to bring the telescope back into spec so that we would receive in-focus pictures.

I think the success of the Hubble Space Telescope repair mission was extraordinarily important to NASA at that time. We needed a success. Hubble was the first satellite to be serviced by people in orbit. The mission was also regarded as a demonstration that we could, in fact, build the space station. Funding and redesign for the space station at that time was always up and down. We needed to show that we could accomplish the goals.

Homer H. Hickam Jr., Neutral Buoyancy Simulator training crew

Hickam arrived at the Manned Spaceflight Center in 1981 as a project engineer for the Spacelab, then volunteered as a scuba diver at the Neutral Buoyancy Simulator at Marshall Space Flight Center.

One day in 1983, I was called to dive on a new mock-up. It was the size of a school bus and it was my first look at the Hubble Space Telescope, an observatory to be launched aboard the space shuttle. Later, I went into the tank wearing the extravehicular mobility unit (EMU) suit to replace some of the HST's modular electronics. I also worked as safety and utility diver for astronauts Kathy Sullivan and Bruce McCandless, who had been assigned to launch the HST from the shuttle cargo bay. The HST was carried aloft in April 1990, and Kathy and Bruce gave the observatory a perfect deployment. Then a chilling announcement was made: The Hubble was fatally flawed. It was nearsighted. Somehow, it had to be fixed.

Another astronaut crew, this one headed up by Story Musgrave, was assigned to fix the Hubble. It didn't take long before it became apparent that the work would require the astronauts to stay outside in their EMU suits for eight hours or more. To allow training for such extended stays, a decision was made to provide nitrox, a gas with a high content of oxygen, to the astronauts in the tank. This meant they would be able to stay for a longer period underwater without worrying about getting decompression sickness. Scuba divers in the tank, however, would still be breathing normal air and be subject to the illness.

To support the long hours of training, every engineer, scientist, technician, secretary, and clerk at MSFC with scuba certification would have to volunteer. Nearly all did. For months, even on weekends and nights, training dives were made again and again. Fatigue became a real problem. Divers were wearing out, but nobody quit. Once, while waiting twenty feet underwater for a piece of hardware to arrive, I draped myself over the HST mock-up and went sound asleep. Astronaut Kathy Thornton woke me up, wondering aloud where her utility diver was. I blinked sleep out of my eyes, spit water out of my mouth, and got back to work. In the end, the sacrifice of the MSFC volunteer divers, the astronauts, and the Goddard Space Flight Center engineers who designed the procedures for the repair all paid off. Hubble was fixed in December 1993, and the universe was opened for all to see as it had never been seen before.

John H. Campbell, program manager

HST went from being the butt of ridicule in talk shows, a "techno-turkey" in one Senator's words, to a national symbol of excellence. How could this be? I never expected that such a thing could happen, certainly not after attending my high-school reunion that awful summer of 1990. My job had only included flight software, yet I was accused of the worst kind of incompetence.

The truth was that the telescope had been built by some of the best people in aerospace. These people were believers in the ultimate capability of the space telescope, and they worked long and hard to do their part to make it so. So what happened with the mirror? Sometimes craftsmen get so confident in their craft and skill they become arrogant, so arrogant that they cannot believe they might be making a mistake. I think that is what happened. If the work had been shoddy, it could never have been corrected. Except for the grossly wrong prescription, the mirror is a fantastic optic. We quickly knew that it could be fixed. The good people didn't leave for some other work; they stayed because they continued to be believers. They believed it could be fixed.

The early adversity bred a culture in the Hubble community that remains today. It is a culture of doing everything possible to get the best science from Hubble, to improve its capabilities and to never waste a minute. The operators of other satellites are happy when they retrieve 99 percent of the data that has been obtained in orbit. The Hubble operators know that the loss of one bit is an embarrassment that must be explained. They have a saying, "A mistake is not an option." A national symbol of excellence emerged from a huge team dedicated to demonstrating that their charge is something that the nation should be proud of. My car has a Hubble Space Telescope license plate.

Below left

Hubble views the Eskimo Nebula, the glowing remains of a dying star. This was the telescope's first image following a December 1999 upgrade.

Technicians inspect

Hubble's primary mirror in 1982. The primary mirror, which weighs nearly a ton, and the secondary mirror make up the Optical Telescope Assembly.

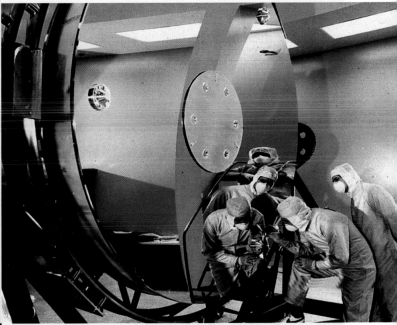

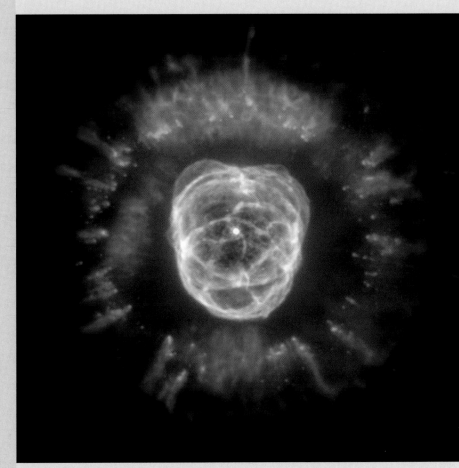

Mark C. Lee, payload commander, STS-82

Lee conducted three space walks totaling nineteen hours and ten minutes during the second servicing mission of the Hubble Space Telescope.

My experience of Hubble goes way back to before it was even launched. It was one of the first things I worked on. I remember I was standing at the end of the Clean Room looking at it, just standing in amazement that we were going to put this instrument up there. I was just awestruck by how really beautiful it was. And then, when you learn about its technical aspects, it's even more amazing. . . . And to be able to work on it, I mean it puts a little pressure on you. You're up there working on something that, if you screw up, you've destroyed one of the most valuable instruments there ever was for mankind. So you never stop thinking about that.

HUBBLE

John Mather, project scientist, Next Generation Space Telescope

I would say that Hubble has been probably the most productive scientific research project in human history. The fraction of scientific discoveries that are reported every year that come from Hubble turns out to be very large— maybe several percent of all science from this one project. And that's a lot. It is just so much more powerful than what we had before.

What it's really enabled us to do is get sharp answers to hard questions. For many decades now, we've been wondering how fast the universe is expanding. We now know that answer within about 10 percent. We have used the telescope to find the most distant objects. And from a public perspective, some of the most photogenic astronomy pictures we've ever taken have come from Hubble.

Steven A. Hawley, mission specialist

The remote manipulator system (RMS) includes an electromechanical arm, known as the Canadarm, which is used in both deployment and maneuvering procedures.

The Hubble, stored in *Discovery*'s payload bay, was deployed from the shuttle during STS-31. I was the robotic arm operator on the mission. Once we got on orbit, my job was to use the arm to grapple the telescope, lift it out of the payload bay, and maneuver it around and secure it while it was checked out and while the various appendages, specifically the solar arrays and antennas, were deployed. Then, once the ground confirmed that everything checked out, I released the telescope and moved the arm out of the way. We then moved the orbiter away from the telescope.

On STS-82, the second servicing mission, we performed the procedure in a reverse order since Hubble was now a free-flying telescope. We approached it, and then I captured it with the remote arm and maneuvered it down into the payload bay where it was latched into a fixture that held it for several days while the guys in the space suits went out and worked on it. While it was in the fixture, I was able to release the arm and use it to help move the spacewalking guys around in a kind of cherry-picking fashion. One of them would actually stand on the end of the arm, and I would move him around to various work sites on the telescope, where he would use the arm as a work platform. That was a lot of fun.

Steven A. Hawley

One of the problems that Hubble was originally designed to tackle was, How big and how old is the universe? The way Hubble addresses this is by looking at a certain type of star that varies its light output over a period of several days. Because these stars exist in the Milky Way, we know that the speed in which they change their light output is a direct measure of how intrinsically bright they are. Therefore, if you see such a star in another galaxy, and you can measure how rapidly its light output changes, you can determine its intrinsic brightness. If you know how bright it intrinsically is, and you can measure how bright it appears to be, you can measure the star's distance from you. . . . The problem has been that you can't find enough of these stars in galaxies that are far enough away to be really interesting in terms of judging the size of the universe. They're just too faint, too small to see from the ground. It was hoped that Hubble would be able to see enough of these stars because of its greater sensitivity and resolution. And, in fact, that's what's happened. Scientists have released some results from their study of these types of stars in faraway galaxies and deduced some numbers that define the age and size of the universe. That is truly amazing.

The Hubble Space Telescope has produced detailed images from incredibly distant galaxies located in what is called the Hubble Deep Field. The light coming from these stellar systems has traveled billions of years to reach Earth, providing a glimpse into the earliest parts of the universe's existence.

The Hubble Deep Field is another great discovery. . . . They pointed Hubble at what they thought was an empty piece of the sky and then took a ten-day exposure to see what was there. What they found was a pictureframe full of galaxies—a phenomenal number of galaxies. They described the size of the sky examined as being as big as a grain of sand held at arm's length, and in that region of space they found thousands of galaxies. . . . Imagine if every place in the sky was like that. It is just mind-boggling.

Opposite	Above
Hubble's Wide Field	Bending light: Hubble's
Planetary Camera 2	image of a massive
caught this spectacular	cluster of galaxies in
view of the Keyhole	the constellation Draco
Nebula in April 1999.	shows the effects of
Hot gas appears as	gravitational lensing.
bright filaments against	Distant galaxies are
dark clouds of dust.	distorted into arcs.

THE COSMIC BACKGROUND EXPLORER

F For years, the Big Bang theory of the beginning of the universe—that it started with a colossal explosion that quickly spread outward in all directions roughly fourteen billion years ago—was plagued by an apparent inconsistency. Theory posed that the explosion had to have been uniform in all directions, yet observation through telescopes turned up lumps: galaxies were unevenly distributed throughout the cosmos.

NASA decided to test the Big Bang theory by sending up a deep-space observatory. COBE, for Cosmic Background Explorer, built by NASA's Goddard Space Flight Center, was launched November 18, 1989. The science team, led by Goddard's John Mather and Rainer Weiss of MIT, used COBE's three instruments to examine the cosmos.

The Diffuse Infrared Background Experiment (DIRBE) team looked for the light from the first galaxies that formed after the Big Bang. It found that the far infrared light from galaxies is just as bright as all the previously known objects combined—a tremendous surprise for astronomers.

The Far Infrared Absolute Spectrophotometer (FIRAS) team measured the temperature of the cosmic microwave background radiation, the presumed remnant of the great explosion itself, and showed that it is the same at all wavelengths. This demonstrated its Big Bang origin and proved there was only one energy source for the Big Bang. Its perfection amazed theorists and experimenters.

The Differential Microwave Radiometer (DMR) team measured hot and cold spots, slight irregularities in the microwave "echo" of the Big Bang, as they were when the universe was three hundred thousand years old. These spots are regions of different density that may have caused the galaxies to clump together. The discovery was hailed by eminent astrophysicist Stephen Hawking as "the scientific discovery of the century, if not of all time."

▼

The first COBE data were so good you couldn't believe them. If the team hadn't provided the output numbers, I could never have believed them.

Philip Morrison, professor of physics, Massachusetts Institute of Technology

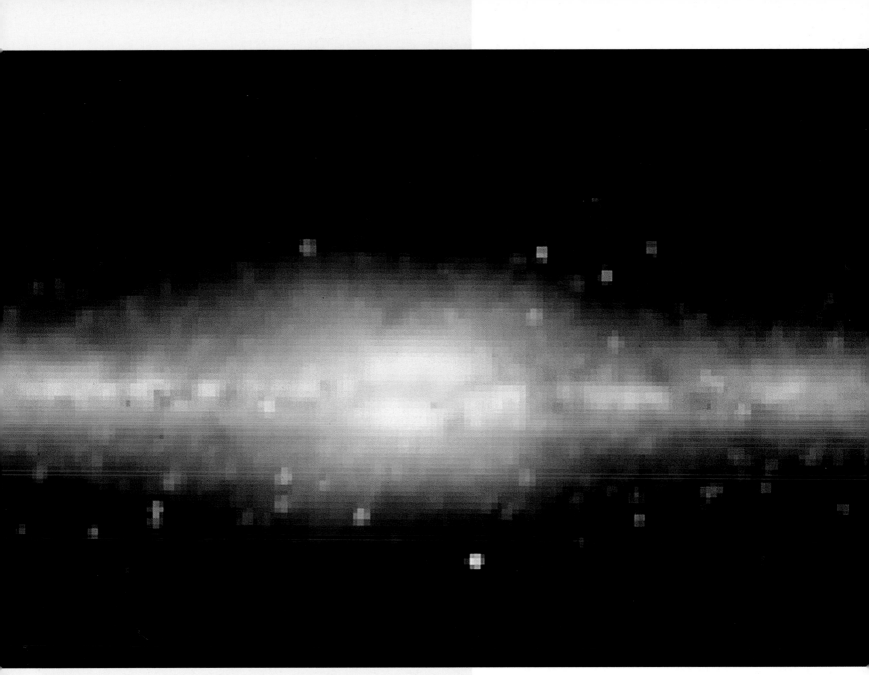

Michael Hauser, principal investigator, Diffuse Infrared Background Experiment (DIRBE)

Searching for the light emitted by all the stars and galaxies formed since the Big Bang often seemed a daunting task. But thanks to the prodigious efforts and talents of many colleagues, I can look back on some wonderful moments. When the last of the old Delta rockets put COBE exactly into the desired orbit, I felt both elated and confident that we were headed for major success. This was followed by a general team euphoria when, a few days later, the cover on the COBE dewar [a liquid-helium cryostat] was ejected and the DIRBE instrument began responding to the infrared light from the solar system and Milky Way galaxy. The near-infrared image of the sky produced by the DIRBE, revealing the bulge of stars at the center of the Galaxy, caught the public's imagination.

It has been an unexpected reward to see the DIRBE results contribute so directly to public understanding of our place in the cosmos. The ultimate high for me occurred when, after twenty-four years of effort, we could report success in finding the long-sought infrared background radiation and begin filling in a new chapter in our understanding of early cosmic evolution.

The central bulge of our Milky Way galaxy is plainly visible in this near-infrared image taken by COBE's Diffuse Infrared Background Experiment (DIRBE). The view covers about 60 degrees of galactic longitude from side to side.

COBE

Philip Morrison, professor of physics, Massachusetts Institute of Technology

They published a curve of the spectrum of the millimeter radiation, a wide-band picture, point-by-point, done by careful measurements. And they got a curve that was the best fit ever obtained to the Planck radiation curve of 1900, which was the first great success of quantum theory. There's no laboratory fit that was anything as good or as wide as that, because you can't set up that kind of a source in the laboratory. The data fit the curve better than the NASA draftsman's circles! When the first audience saw that slide, they just got up spontaneously and clapped and cheered.

How cold is the dust between the stars? COBE's Far Infrared Absolute Spectrophotometer (FIRAS) saw bright regions as slightly warmer (-411 degrees Fahrenheit) than dark regions.

John Mather, project scientist

Mather was principal investigator for the Far Infrared Absolute Spectrophotometer (FIRAS), which measured the detectable radiation afterglow of the Big Bang. Theoretical models had predicted that the expansion of the universe cooled this radiation from its initial trillions of degrees to near the measured 2.728 degrees above absolute zero.

The first result to come out was from my experiment. It agreed perfectly with this theoretical model within about a percent of accuracy, which is all we could get from [the first] nine minutes of data. After we analyzed all the data that we took over the course of ten months, we were able to improve this accuracy to fifty parts per million, which was way beyond what we'd ever dared to promise. But the upshot was that the Big Bang prediction was confirmed as well as you could imagine.

To tell the truth, I didn't expect that we would have anything very different. However, it was only a week or so after launch that we were able to start processing some of the data and get an inkling that this was so. I still have somewhere an autographed spectrum of the first results from FIRAS. And all of the team members who were processing the data late at night produced this thing and autographed it, and gave it to me the next morning. They were the people who saw it first.

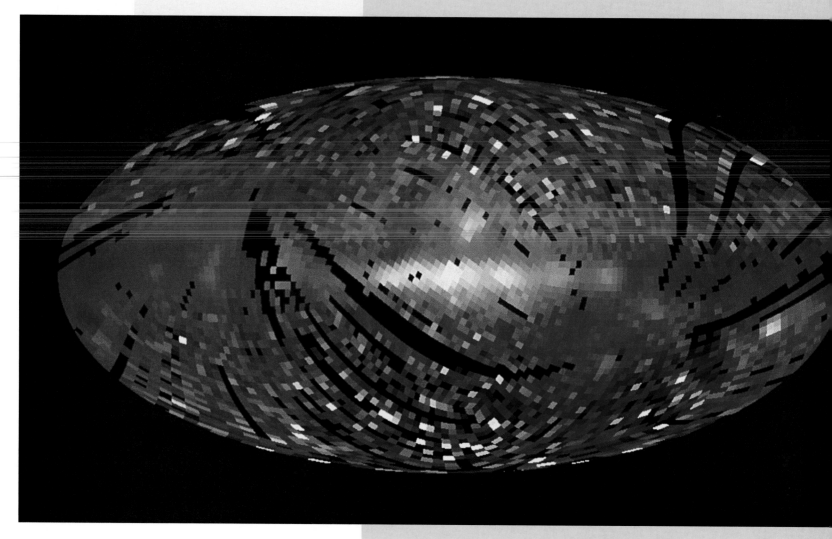

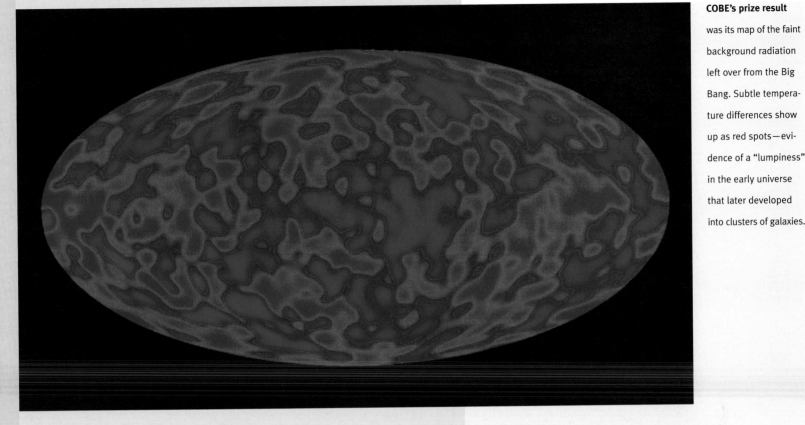

COBE's prize result

was its map of the faint

background radiation

left over from the Big

Bang. Subtle tempera-

ture differences show

up as red spots—evi-

dence of a "lumpiness"

in the early universe

that later developed

into clusters of galaxies.

John Mather

The differential microwave radiometer (DMR) detected temperature fluctuations in the cosmic microwave background on a scale of thirty millionths of a degree. These minute differences represent density variations in the early universe necessary to explain the observed distribution of galaxies.

The first moment where the whole science team knew that something was up was when [UCLA astronomer] Ned Wright brought a collection of 35 millimeter slides to an evening meeting of the science team at [NASA program scientist] Nancy Boggess's house. He showed these slides and said "I think this is real. I think we found something." And most of the rest of the team said, "Well, maybe so, but we don't think it's ready to show anyone, because it's only you saying it. So we've got to go do it again."

It took us maybe six or eight months to finish that. While the suspense was building, some people were letting out a few leaks about what we might be finding. But we tried very hard to keep a tight lid on any public announcement of anything. We all were sworn to secrecy.

We were all overwhelmed with the difficulty of the software problem [writing data analysis programs], so we knew that even if Ned said it was true, it might still be wrong. I would say Ned's one of the smartest people I've ever known, but nevertheless, it doesn't mean that what he said was true. . . .

Dave Wilkinson of Princeton, another team member, was the most worried and the most difficult to convince about these things. And since we all had a solemn agreement that we would all agree before we went public, we really tried hard to convince him. So people made presentations and argued with him privately and in public for quite a while before we could say "Yeah, okay, maybe even Dave agrees with it now." It's one of my principles that you need people around who will tell you the truth. You need your own internal skeptics.

George F. Smoot, principal investigator, Differential Microwave Radiometer

Two times stand out when I think about COBE: Watching the bright, noisy rocket take off carrying the instrument on which I—and many others—had spent so many years and so much effort preparing and testing. Then not long afterwards, the DMR instrument was turned on and in the first half minute we saw a signal, then a signal as the moon swept by, and right afterwards the internal calibration source. Over the phone I heard the excited cheer from the majority of the team in the science operations room and knew the instrument was working. It was now up to us to find out what Nature supplied.

After much work there was the thrill of knowing that the instrument had made a discovery of literally cosmic significance. That discovery set off a flurry of public interest and more importantly a large long-term effort directed at using the newly discovered features to learn about the creation and evolution of the Universe. The thrill first came slowly. I worked to deny it until we could carefully check that it was a real cosmic signal and not something from the instrument or other error. Then I felt a form of anesthetized euphoria and excitement as we prepared to make the findings public, finally a release of results and feelings.

MAGELLAN UNVEILS VENUS

ysterious Venus, with her cloud-shrouded atmosphere, was the first planet visited by a spacecraft from Earth, when Mariner 2 came calling in August 1962. Then she turned into a heartbreaker whose fatal attraction wrecked a string of Soviet spacecraft until Venera 4 arrived in late 1967, closely followed by Mariner 5. Other spacecraft paid visits, but the gaps in knowledge remained enormous.

As the years passed, scientific interest in the planet grew, largely because its surface, the temperature of which is hot enough to melt lead, lies under a 97-percent carbon-dioxide atmosphere that produces an extreme example of the greenhouse effect that may threaten Earth.

The Venus Orbiting Imaging Radar mission, eventually renamed Magellan, was launched from the shuttle *Atlantis* on May 4, 1989, and in September 1990 began systematic radar-mapping of nearly the entire planet from a polar orbit with a resolution of over three hundred feet. Magellan also collected data on Venus's gravity field, its efforts later improved by an as-yet-untried technique called aerobraking, which resulted in a lower, more circularized orbit, as low as 112 miles, for the spacecraft.

Much of what the mission turned up astounded geologists. There were pancake-shaped lava domes arranged in straight lines; sand blown into the shape of giant horseshoes; clear evidence of volcanic resurfacing; lava channels thousands of miles long; and so much more atmospheric, geophysical, and geochemical information to fill fat books.

In October 1994, Magellan was deliberately plunged into Venus's atmosphere, completing one of the great missions of twentieth-century exploration.

▼

Magellan was a crucially important mission for us to get closure on our understanding of terrestrial comparative planetology. . . . Venus may turn out to be a laboratory for the study of early Earth history.

James Head, Magellan scientist

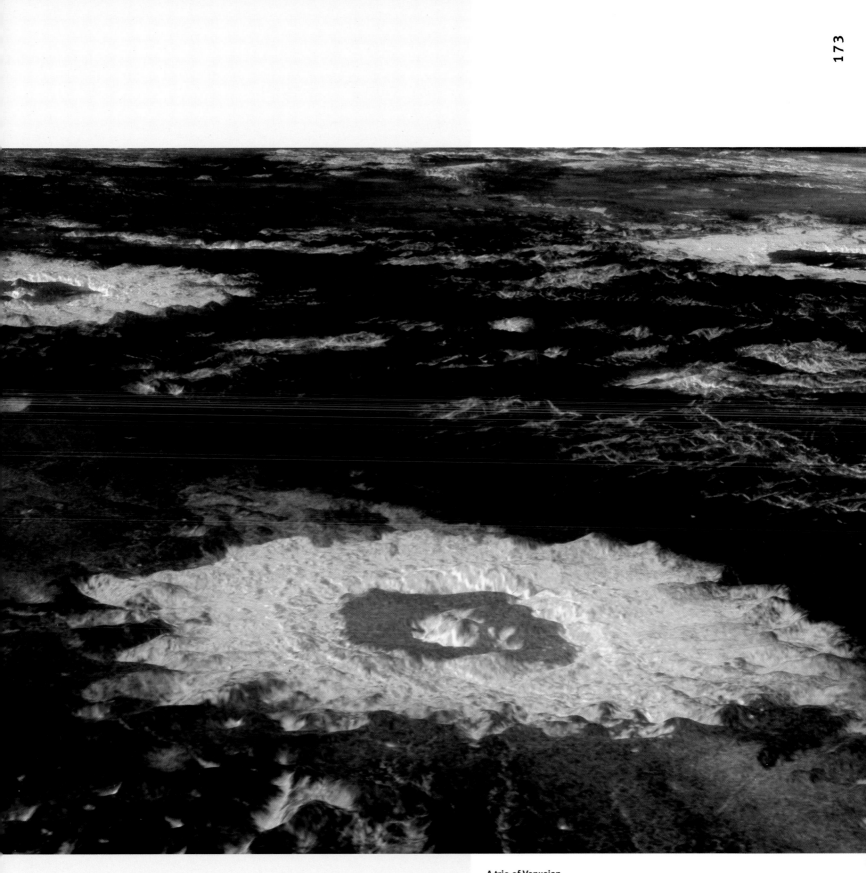

A trio of Venusian craters, as seen by Magellan's orbiting radar imager. The nearest crater, named Howe, is 23 miles across. Its central peak and ring of ejected material are typical of impact scars.

MAGELLAN

James Head, Magellan scientist

At the time of Magellan's launch on May 4, 1989, Head was in a taxicab heading into New York City from the airport.

The news came on the taxicab radio, and, at the time, I didn't know what had happened with Magellan. The first thing I heard was something about Lyle Alzedo, a football player for the New York Giants, and how he had just broken up with some Scandinavian actress. The second story was about Magellan having launched safely and on its way to Venus. It was treated kind of like an afterthought to the "big story." I sat there thinking, and then, since I usually try to talk to cab drivers because they're always interesting in one way or another, I said to this guy, "So that Alzedo, you could see that coming like a freight train, eh?" And he answered, "What? Oh, I'm sorry, I was thinking about what Magellan was going to do on Venus." And I just died. . . . To me, the funny part was not that I could tell him about Magellan, but that I had had this stereotype that he wouldn't give a damn about the mission and so had broken into this typical "everyman" kind of thing—Oh, those guys and those actresses. It was too funny. But we did have a nice conversation.

Alexander "Sasha" Basilevsky, Soviet guest investigator

Magellan's success was built on missions that preceded it, many of them Russian. Basilevsky, who had worked on prior Soviet Venera missions, was part of a Russian-American scientific collaboration to interpret mission data.

Venera's panoramas gave us a close-up view of the ground. Looking from the lander's ground-level view, you could see soil and rocks but only at a distance of twenty to fifty meters [twenty-two to fifty-five yards]. What Magellan gave us was a regional background. Now we could see what was going on in this place for kilometers and discover the kind of regional geological units [individual features] that were around. Also, most of the Venera landings measured the chemical composition of the surface. Now with Magellan, we could correlate the chemical composition of the surface with the type of geological unit. That geological unit was interpreted and understood based on Magellan images. That was really exciting.

Douglas G. Griffith, project manager

We knew a little bit about the type of data we were going to see, but nothing near the resolution we were going to see with Magellan. Russian landers had taken a few pictures, so we had an idea of what the surface would look like. What we didn't know was the whole topography of Venus that was going to unfold, such as the immense number of impact craters and meteorites, over a thousand of them. There was a whole raft of volcanoes, now dormant, that once were major expellers of lava. Then we saw all of those lava channels that went for thousands of miles, which meant that the surface at one time on Venus must have been extremely active and completely resurfaced by volcanic flow. That told us a lot about the age of the last major volcanic eruptions on Venus—maybe five hundred million years for major volcanic resurfacing.

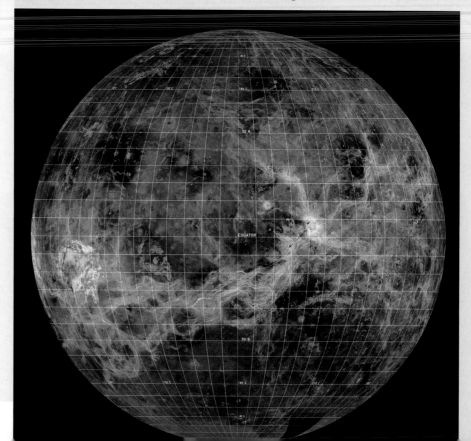

This global view of the surface of Venus is centered at 180 degrees longitutde. Over the course of four years, the Magellan spacecraft mapped more than 98 percent of the planet's surface, showing details as small as 100 yards across.

View of Venus centered

at 180 degrees east

longitude. Magellan's

radar pictured the sur-

face in thin strips,

patiently building up a

global map.

Below

Odd, pancake-like

features appear in

the Magellan radar

images—the product

of viscous lava flows

on a world so hellish

that lead would melt

on its surface.

James Head

*Magellan's radar mapper imaged a single swath of Venus's surface at a time,
radioing data back to Earth at the end of each three-hour, fifteen-minute orbit.*

We'd take a strip of data and look off to the side for the image and
then straight down for the altimetry data. It would take two or three days to
get adjacent orbits. The data would come in on about eight-inch-wide by fif-
teen-foot-long prints, which resembled very bad Xeroxes; they were almost
wet-looking. At the end of the day, the engineers and the process people
would come in with a shoebox full of these rolls. They'd open the door to the
science analysis room and throw this box in, and we'd have this feeding frenzy.
We'd get the prints, roll them out, and put them up on a wall. We'd already
have stuff up from previous days. We'd mosaic pieces together and then
arrange them in quads. The picture would begin to emerge. We had issues
about whether there were plate tectonics or not. We had issues about what the
crater density was. Ideas were being tested. Every day was a constant discovery in that sense. It was unbelievable. It was fun. We had a very good team.

Douglas G. Griffith

Toward the end of the mission, we were within about two or three
weeks of running out of power. During orbit, solar panels experienced great
extremes of heating and cooling. Solar cells and solar panels flexed, electrical
connections broke. This was expected. So we decided, OK, let's do one last
experiment. Let's put the spacecraft into a suicide dive into the atmosphere, a
really high-torque windmill experiment to get as much data out of the mission
as we could. But surprisingly, after that began on October 11, 1994, it took several orbits for the spacecraft to get far enough into the atmosphere to send its
last signal on October 12.

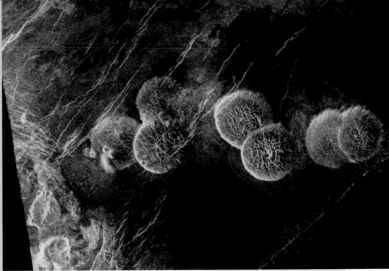

Douglas G. Griffith

We always used to think of Venus as a sister planet
to Earth because of its similar size. It has clouds, so we figured it had an atmosphere of some kind. Until we were able
to pierce through its clouds, we had speculated that the
planet had no liquid on its surface. What happened to
Venus, being that it was closer to the Sun than Earth, was
apparently a runaway greenhouse effect. All of the liquids
that might have originally been on the surface of Venus
evaporated, causing the buildup of this tremendous atmosphere. . . . Everybody theorized for years that Venus was a
lot more like Earth. At one time it probably was, but that was
probably billions of years ago.

GALILEO AT JUPITER

Galileo was launched on the shuttle *Atlantis* on October 19, 1989. Its mission, following the Pioneer and Voyager flybys, was to study Jupiter's atmosphere, satellites, and surrounding magnetosphere.

The two-and-a-half-ton spacecraft suffered an inauspicious start, marked by a seven-year launch delay due to political and technical problems and the 1986 *Challenger* accident. In addition, while en route to Jupiter, an umbrella-like main communications antenna refused to open completely and a data tape recorder became stuck. Despite these setbacks, Galileo completed three planetary and two asteroid encounters in the course of its gravity-assisted flight and, on December 7, 1995, swung into orbit around Jupiter with faultless precision, becoming the first to orbit the gas giant and sample its atmosphere in situ.

Galileo's entry probe, which was released five months earlier, slammed into the Jovian atmosphere at over one hundred thousand miles an hour. During the just under fifty-eight minutes between its parachute deployment and its failure and meltdown, it sent on-the-spot atmospheric, radiation, and other data to the Galileo orbiter for relay back to Earth.

Well into 2000, the supposedly "doomed" spacecraft scrupulously photographed and analyzed in unprecedented detail not only Jupiter, including its rings and cloud bands, but also its four major moons: Io, Europa, Ganymede, and Callisto. Gravitational and magnetic data suggest that all but Callisto have planetlike cores of iron or rock. Io was seen to have eighty or more volcanoes. But the discovery that stole the show was on Europa, where Galileo turned up evidence of a frozen-over, deepwater ocean that could contain life.

▼

We presented a copy of our first images of Europa from the Galileo spacecraft to Pope John Paul II, who probably gave the most succinct summary yet of these data: "Wow."

Torrence V. Johnson, project scientist

A 3-D visualization of Jupiter's atmosphere near an equatorial "hotspot." The Galileo probe descended through such a region on December 7, 1995.

GALILEO

Torrence V. Johnson, project scientist

When the high-gain antenna on Galileo wouldn't open, it meant that although we might get to Jupiter, we wouldn't be able to do anything once we got there. We were left with the secondary antenna system. We could deliver the probes successfully, but we would have very low data rates; that got pretty close to a mission failure.

It was a tense time. The engineers didn't want to give up on getting the high-gain antenna open. They worked very hard to devise all kinds of clever ways to heat and cool the spacecraft to see if they could pop the antenna open. Finally, we had a meeting and said, "Look, it's wonderful that you guys are trying so hard to open that antenna, but we all have to recognize that the probability of that working is pretty nil."

So we started thinking of other options. That led us to some ideas that allowed us to recover from the antenna failure. We improved the sensitivity of our antennas on the earth, and we worked to compress data so we didn't have to transmit as many bits. We had to, in effect, do "brain surgery" on the spacecraft to change the way its computers were working. It was a tremendous task, and the fact that it all worked as smoothly as it did still amazes me.

Merton E. Davies, imaging team

Jupiter is a very exciting planet. It's the largest one in the solar system. It has some of the largest satellites in the solar system. Of course, the four large Galilean satellites are the ones that get most of the attention, and the reason is that each tells a unique story. Europa, for example, has a surface that is very fragmented and cracked, and as you study it, you can see a lot of motion in these blocks of ice. And you can see that they have moved relative to each other. At least one explanation is that underneath them is an ocean similar to what we have in the North Pole. . . . And if there's water, you always think of the possibility of life.

Above

The four large moons of Jupiter discovered by astronomer Galileo in 1610—as photographed by his namesake spacecraft nearly four centuries later. From left: Io, Europa, Ganymede, and Callisto.

Richard Young, probe project scientist

The tension in the control room I was in at JPL was immense just prior to the predicted time of entry of the Galileo probe into Jupiter's atmosphere, December 7, 1995. We had not been in communication with it for five months; to conserve its batteries, the probe had been kept unpowered since its release from the Galileo orbiter five months earlier and some fifty million miles from Jupiter. Other than our predictions of where it should be, the condition and position of the probe were completely unknown. We knew that the probe entry into Jupiter's atmosphere would be the most difficult entry ever attempted in terms of speed (106,000 mph) and heating loads. Furthermore, it had to enter the atmosphere within a very narrow corridor: Only a 1.5-degree error, plus or minus, could cause loss of the probe. What were the chances it would survive?

When we received confirmation that the orbiter had locked onto the probe signal after entry—the first telemetry from the probe in five months—there were tears in people's eyes. After all, some of the people in that room had worked on the project for almost twenty years, and we had all waited through a trajectory to Jupiter that took over six years from launch to encounter with Jupiter.

Torrence V. Johnson

If Europa has an ocean, its volume would be more than twice the volume of the Earth's oceans. Conditions in such an ocean, overlying a possibly volcanic seafloor, could be similar to those prevailing in seafloor hydrothermal vent systems on the earth, where we now know that heat-resistant primitive life-forms can exist and may even have originated. Would similar conditions lead to life on a place like Europa? We'd sure like to know the answer!

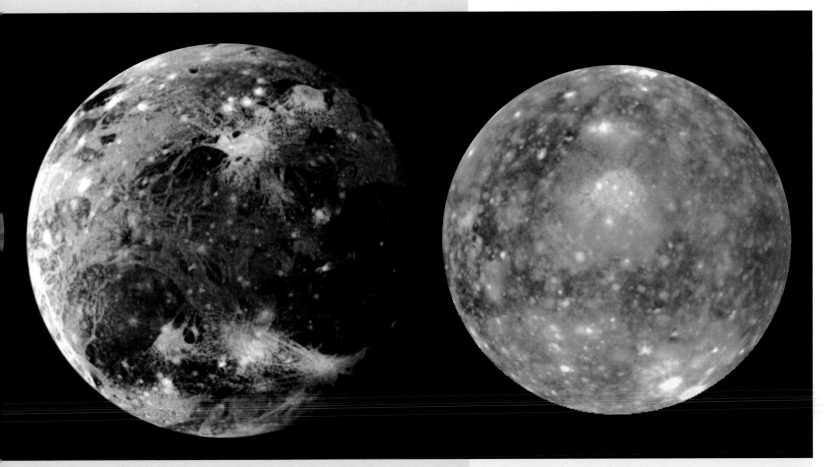

Margaret G. Kivelson, principal investigator, magnetometer

Galileo's unique trajectory allowed its magnetometer, which has changed our view of the magnetism of Jupiter and its moons, to make a series of other interesting measurements en route to the giant planet.

Indeed, the six years of Galileo's interplanetary travel to Jupiter were a bonus because we got outstanding data from its Venus flyby in 1990, which contributed to the study of planetary bow shocks—the disturbance that stands upstream of a body in the solar wind. Then we had two flybys of Earth and were able to milk the data for very interesting scientific results. This was largely because Galileo's orbit was so different from the orbits of spacecraft that had previously explored Earth's space environment. In addition, Galileo performed two close flybys of asteroids and, to our great pleasure, there were significant signatures in the data that we acquired that opened up new ideas about how an asteroid might interact with the solar wind. So, it was not without scientific interest to acquire the data on the way to Jupiter; we were very happy that we were allowed to do that. . . . As for the magnetometer's performance at Jupiter, I just couldn't have imagined that we would have such a trouble-free instrument for such a prolonged mission with the presence of so much punishing radiation. Other than losing data from one passby of Europa, during which an energetic cosmic ray knocked out an operating control component of the magnetometer, we've had no troubles at all with the instrument. Given the kind of environment we've been functioning in, that's absolutely remarkable.

Hot stuff on Io: Dark plumes from the volcano

Prometheus appear regularly in Galileo photographs of the solar system's most active moon.

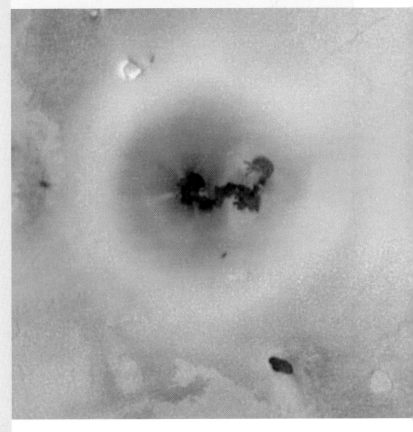

SERVANTS IN SPACE

I f the point of having outside help is to make one's life easier and more efficient, then civilization as a whole can be considered to be better off for having its own type of outside help—far outside, in fact.

Although no space-borne machine can fix breakfast, iron shirts, or drive the kids to school, many of them are, in fact, robotic "servants" that have made the lives of uncounted millions of people easier and more productive than could have been dreamed of a century ago.

Consider the far-reaching benefits that satellite technology has brought us: News that once took weeks to travel across an ocean is now relayed almost instantly by communications satellites. Farmers previously vulnerable to unpredicted frost, storms, floods, and other weather hazards can now receive ample warning through a constant stream of forecast data from meteorological satellites. Civil engineers can survey the best terrain for mountain roads, railroad bridges, canals, and other routes in a fraction of the time it once took, thanks to topographic maps made by remote-sensing satellites. Ship captains sailing anything from ocean liners to sailboats can stay on track (and tack) with the help of a fleet of navigation satellites. Backpackers, drivers, and other travelers can know exactly where they are at any given time through the guiding satellite fleet that makes up the

Hurricane Hugo bears down on the South Carolina coast, September 21, 1989. The massive category-four hurricane is caught in a computer-enhanced view based on data from the GOES-8 weather satellite.

Global Positioning System. And finally, ranchers, foresters, and those who track population densities and movement can use highly detailed maps from remote-sensing satellites to track everything from bison and cattle movement to redwood forest growth to the expansions of cities from Los Angeles to Nairobi.

The three most obvious applications for high-altitude, Earth-orbiting machines—weather prediction, communication, and "remote sensing" to monitor the land, sea, and air—were envisioned even before the space age began.

The cloud or storm systems, including hurricanes, we see moving across the television screen are imaged by the descendants of two experimental weather satellites: TIROS (Television Infrared Observation Satellite) and NIMBUS. TIROS 1, the first in a series, went into orbit in April 1960 and sent down twenty-three thousand pictures during its seventy-eight-day life. The TIROS series was followed by seven NIMBUS satellites, which carried still cameras, television cameras, and infrared radiometers for night imaging. Since TIROS orbited at four hundred miles and NIMBUS at six hundred, the more advanced of the two satellite types carried weather prediction to new heights in both senses of the term.

TIROS and NIMBUS were replaced by a series of constantly improved spacecraft. ESSA satellites, followed by NOAA satellites, which flew from 1966 well into the 1990s, extended the altitude to nine hundred miles but could send down pictures only once a day. That changed with the launch of GOES (Geostationary Operational Environmental Satellite) in May 1974. As its name suggests, GOES satellites and their successors are parked in geostationary orbits out at 22,300 miles above the equator, where they keep pace with the rotation of Earth, thereby providing continuous weather coverage.

Fifteen years after the science-fiction writer Sir Arthur C. Clarke suggested using spacecraft to connect a global communications system, a one-hundred-foot-in-diameter Mylar balloon called Echo 1 was put into orbit and used to ricochet radio signals from one place to another. That launch in August 1960 was followed by an endless number of communications satellite successors, many of which have been built and launched by ham-radio enthusiasts. OSCAR 1, the first "hamsat," was launched in December 1961, showing that it didn't take the amateurs long to figure out how satellites could expand their world. On July 10, 1962, NASA launched AT&T's Telstar 1, the first "comsat" that received, amplified, and retransmitted signals. Early Bird, also known as Intelsat 1, went up in April 1965 as the first commercial communications satellite.

Eleven years later, communications satellites went into use as relays for commercial television, allowing events such as the Olympics to be seen in "real time."

One of the most important gifts of the space age has been the ability to look at Earth from a great distance: in effect, to step back and examine it in large chunks for a variety of reasons. The first civilian remote-sensing spacecraft, called an Earth Resources Technology Satellite (ERTS) and then simply Landsat, went up in July 1972 and started a revolution in planetary management and understanding. It and its successors, working in circular orbits some five hundred and seventy miles up, carried television cameras and multispectral scanners that measure reflected light from the ground—radiation—in various wavelengths.

The pictures that have streamed down since Landsat 1 went into operation have had a profound impact. They have spotted new earthquake fault lines in

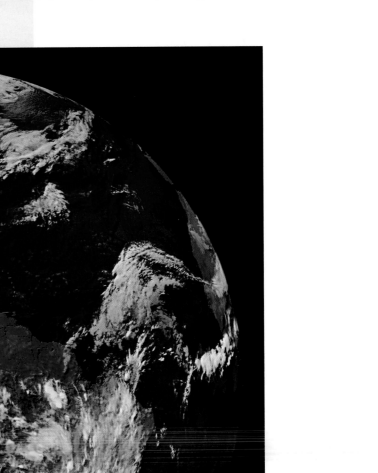

Earth as seen by GOES-8 in January 2000. The venerable GOES (Geo-stationary Operational Environmental Satellite) line of observers has provided continuous weather coverage from high orbit since 1974. GOES-11, the latest in the series, was launched in May 2000.

California; tracked vegetation density (and therefore mosquitoes) in the Rift Valley to help stop the spread of a deadly virus; and in 1986, spotted radioactive gas pouring out of the stricken reactor at Chernobyl in Russia.

But the greatest value of remote-sensing capability is seen in the workaday world of resource measurement in agriculture, forestry, land use and mapping, geology, water use, oceanography and marine resources, and the environment in general. Landsat imagery is used, for example, to monitor surface mining and reclamation, map and check on water pollution, spot air pollution and measure its effects, calculate the effects of such natural disasters as hurricanes and earthquakes, and measure the effects of human-caused damage such as defoliation and acid rain.

All of these satellites, as well as a fleet of others that collect biological, physical, and other data in the service of humankind, are truly civilization's servants.

COMMUNICATIONS SATELLITES

Radio waves generally travel in straight lines and therefore cannot bend around the horizon. One of the earliest and most obvious uses for the altitude advantage that space provided, therefore, was communication: Great height could dramatically increase the distance and speed a signal could travel.

From the beginning, civilian communications satellites for radio, telephone, and television would be developed and operated by the private sector for profit. Although a couple of passive communication satellites, which simply reflected radio signals back to Earth, went up earlier, modern satellite communication really began in 1962. On July 10, AT&T Bell Laboratories' Telstar 1, the first active real-time communications satellite, was launched by NASA. Rather than bounce weakened signals, it picked them up and retransmitted them. In December 1962, RCA's Relay 1 went up with twelve telephone channels and the first television channel. These low orbiters were joined in 1963 by Syncom 2, the first satellite to broadcast television pictures from geosynchronous orbit. The announcement "live via satellite," now long vanished, became the symbol of cutting-edge technology.

The Communications Satellite Corporation (COMSAT), created by President John F. Kennedy to establish a worldwide satellite network for peace and understanding, led to the International Telecommunications Satellite Organization, or Intelsat, which started in 1964. It is owned by member nations, with an increasing share going to developing countries, and it launches its own advanced satellites.

Whole fleets of smaller privately owned communications satellites are on the horizon (or at any rate over it) that promise to further speed communication as part of the planet's central nervous system.

So, during the second year of ATS-6's life, this satellite experiment was performed. It was called SITE, the "Satellite Instructional Television Experiment." They would broadcast an hour and a half worth of television every evening when the Sun went down, and these television sets were placed in villages that perhaps had only one electrical outlet in the village. I attended a few of these, where the TV was plugged into that only outlet and the villagers came and sat to watch for that short time. The programs were generally a half hour of news, a half hour of instructional material having to do with a lot of things—health, agriculture, what have you—and then a half-hour's worth of entertainment.

This went on for a year. Near the end of the program, I visited India. I went from village to village in a Jeep, watching some of these programs take place. We would ask the people what they got out of the program. I would get a lot of standard answers—it was entertaining, we learned a lot. I asked one gentleman who looked rather aged, about fifty or thereabouts, what he got out of the television program. He was very proud. He said, "You know what I got out of this television? I saw a fat cow!" He had not known a cow was supposed to be fat.

Leonard Jaffe, director, communications satellite program, NASA Headquarters

There's no question in my mind that communications of all sorts, but particularly by satellites, have changed the world in all of its aspects. We are a single world today, and what's going on in one part of the world affects everybody in the world. I've got a great anecdote that I'd like to relate. This has to do with the ATS-6 satellite. ATS-6 was an experimental satellite that NASA launched into geosynchronous orbit, and it was the first satellite that had enough power on it to radiate signals that could be received by reasonably sized television sets on the ground. It was launched in 1974.

The government of India came to the United States, and to NASA, and asked if they could develop a cooperative program in which we would move the satellite over so that it could be seen by the Indian continent, and they would mount a rather ambitious educational experiment for the country. We agreed to move the satellite in the second year of its operation. Meanwhile, India, which had not had any television manufacturing capability, built several thousand television sets, largely with the help of a lot of college graduates (wired by hand), and these were distributed throughout several thousand villages. They also developed an ability to produce programming, educational programming, that would attract the population.

In addition to conducting technological experiments, the ATS-6 became the world's first educational satellite and transmitted programming to remote parts of the United States, India, and other countries.

Above left

NASA's Applications Technology Satellite-6 (ATS-6), the most complex, versatile, and powerful communications research satellite of its time, undergoes final test prior to shipment to Cape Canaveral, Florida, for launch on May 30, 1974.

TAKING THE GLOBAL VIEW

L ong before the first breathtaking image of Earth was sent back by Apollo 8 astronauts in 1968, NASA realized the potential of space travel as a tool for studying the planet. Relying first on imagery supplied by aircraft and from manned spacecraft, NASA began monitoring food production, environmental change, and population movement from space.

In 1972, the agency launched the first Earth Resources Technology Satellite (ERTS-1). Specifically designed to view the changing Earth, it later became known as Landsat 1. In the years that followed, improved Landsat successors, such as Landsat 7, launched April 15, 1999, and the sophisticated remote-sensing instruments that currently fall under NASA's Earth Science Enterprise have monitored ozone depletion, tracked and predicted global climate and local weather phenomena, mapped the globe in greater detail than ever before, and examined the state of the oceans.

The Geostationary Operational Environmental Satellite system, or GOES, for example, provides the National Weather Service with continuous information for weather forecasting, particularly of severe storms. Other Earth-monitoring systems are shuttle-based. In early 2000, shuttle astronauts carried out the Shuttle Radar Topography Mission (SRTM), gathering data for an unprecedented high-resolution topographical map of the planet. The oceans are monitored through a range of ongoing programs. TOPEX/Poseidon observes global sea levels and ocean circulation and their impact on the atmosphere. Launched in 1992, it tracks oceanic phenomena such as El Niño and La Niña and their impact on global climate.

The Sea-viewing Wide Field-of-view Sensor (SeaWiFS) project monitors the changing color of the world's oceans, which offers critical information about the microscopic marine plants on which all ocean life depends.

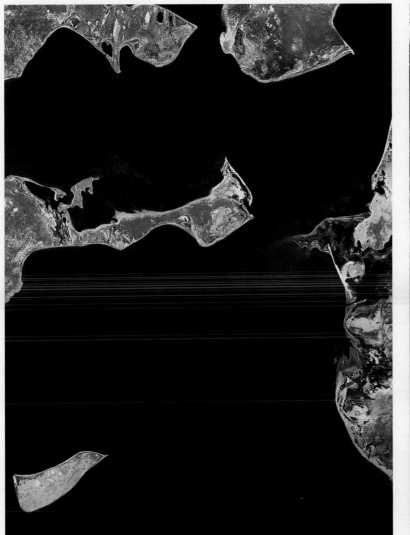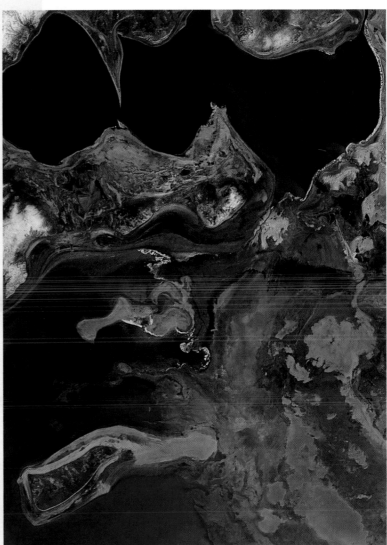

R. J. Thompson, Landsat 7 program manager

One of the values of Landsat is that it illustrates a practical application of change detection—of monitoring change detection anywhere in the world. We can monitor how this planet has changed and get some clue as to why it has changed. For example, in the case of the Aral Sea, Kazakhstan, this very large body of water has greatly reduced in size, and it's a cause-and-effect situation. There was a cause for why that lake shrunk. When somebody started to investigate it, they found out that it had to do with upstream farming practices, among other things. So here, an activity that was completely divorced from and unrelated to this particular body of water has had dramatic consequences. People had begun to die because of the pollution and the rise in saline levels of the only water available to them.

The shrinking of a sea. Photographs of an area of the Aral Sea in central Asia taken by Landsat 1 in 1973 (left) and Landsat 7 in 1999 show the dramatic encroachment of land as the sea has dried up, its water siphoned off for irrigation.

GLOBAL VIEW

R. J. Thompson, Landsat 7 program manager

We have about a dozen international receiving stations around the world signed up for the Landsat 7 program. Countries pay a fee to acquire the data from us and then use the data for their own purposes. We had a meeting in Neustrelitz, Germany, in September 1999, and it was rewarding to sit in a room with nonpolitical—basically apolitical—representatives who were purely science oriented and interested in the technology. To listen to them talk with so much enthusiasm about the availability of this kind of data for their area. . . . I mean, these are countries that can't afford to fly satellites themselves and would never see this technology were it not for the fact that we're making it available. There is such global enthusiasm for the fact we have a technology that we can all take advantage of and, if we use it correctly, will allow us all to learn more about each other and break down some cultural and political barriers. We had people at the meeting from countries that typically won't talk to each other at the political level, and we were just having a wonderful time. There was China, Indonesia, Thailand, South Korea, a representative from Russia and, of course, all of the European countries, then South America, Africa, etc. We're all interested in the same thing.

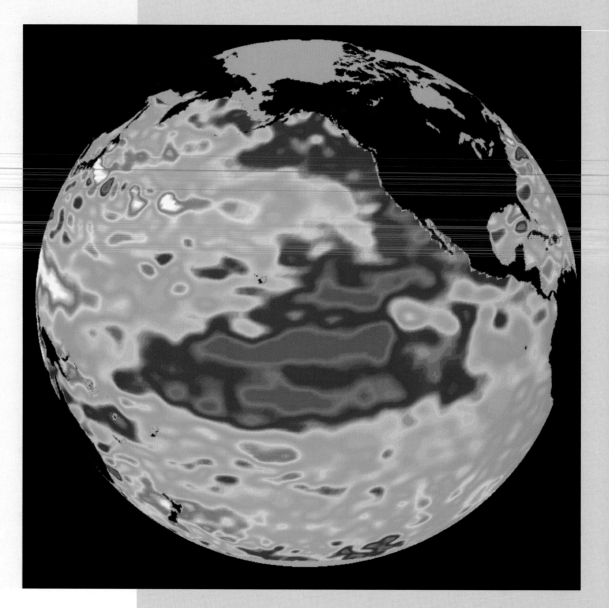

James E. Hansen, NASA Goddard Institute for Space Studies

TOPEX measures sea level to an incredible accuracy, on the order of millimeters. This is useful for understanding sea-level change, which is one of the most important impacts of global climate change. Sea level is expected to rise, and is already rising, because of global warming. This warming causes ocean water to expand, and adds more water to the ocean by melting mountain glaciers and the ice sheets that form Antarctica and Greenland. Sea level rose about fifteen centimeters [about six inches] in the last one hundred years. Will it rise a few tens of centimeters this century, which would be a nuisance, or a meter or more, which would be a disaster?

The joint NASA and French space agency's TOPEX/Poseidon satellite has become a powerful tool for tracking El Niño and La Niña, which periodically wreak havoc with North American weather. The satellite measures sea surface height and temperature, with cold water appearing in blue and purple and warmer water appearing in red. Below left image was taken in May 1999; below right image shows dramatic changes three months later.

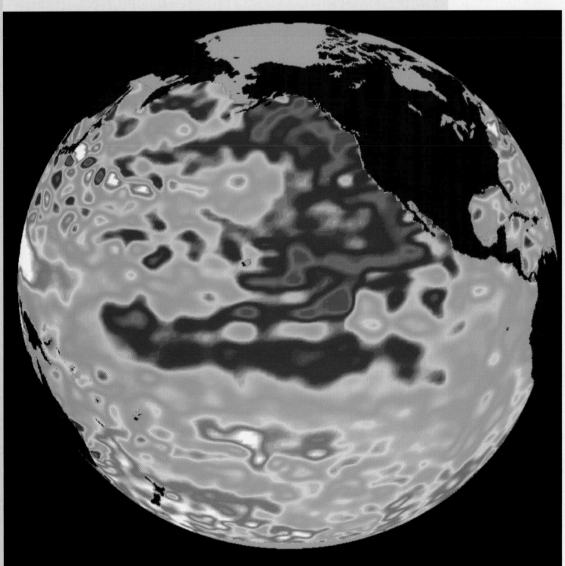

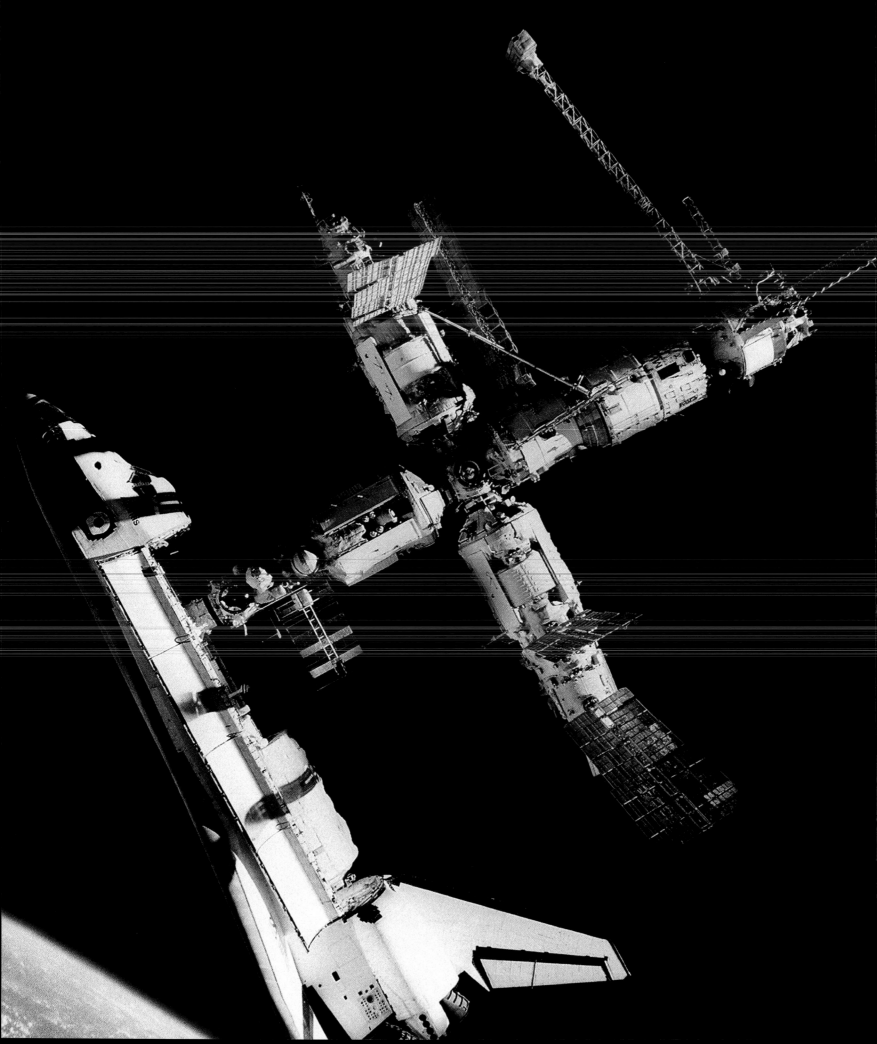

OUTPOSTS IN SPACE

I n October 1869—almost exactly a hundred years before Apollo 11 went to the Moon—the *Atlantic Monthly* ran the first in a series of articles by Edward Everett Hale that described a man-made moon. "The Brick Moon," a proposed sixty-yard contraption that held thirty-seven crewmen, was supposed to help ships navigate on the high seas.

No one who has thought seriously about sending humans to space has considered short orbital missions as an end in themselves. The idea has been that once people reached space, they would remain there for long periods. Space stations in various forms have been designed from the outset to allow this, as well as fulfill two basic functions: to serve Earth continuously in several capacities (including observation) and to provide a platform from which to begin humankind's migration to the Moon, Mars, and beyond.

The idea of humans inhabiting an orbiting outpost as a first step to the migration to space was central to visionaries such as Russian inventor and rocket expert Konstantin Tsiolkovsky, German astronautical pioneer Hermann Oberth, and noted rocket engineer Wernher von Braun.

Von Braun's two-hundred-fifty-foot-in-diameter, wheel-shaped station, described in *Collier's* magazine in 1952, would have held eighty people and created its own gravity by spinning in an orbit a little more than a thousand miles high. "Life will be cramped and complicated for space dwellers," wrote Willy Ley, one of von Braun's collaborators. "They will exist under conditions comparable to those in a modern submarine." The wheel, which was a flying hotel compared to later designs, never made it

Atlantis docks with the Russian Mir space station. Together they formed the single largest spacecraft ever placed in orbit, with a total mass of almost half a million pounds.

off the drawing board because it was far ahead of its time and fabulously expensive.

Having lost the race to the Moon in 1969, the Soviet Union decided to lead the way in manned orbiting outposts and launched the world's first space station, Salyut (salute) two years later. (The "salute" was to Yuri Gagarin, whose historic flight had been made almost exactly ten years earlier.) Salyut 1 was followed by four other first-generation stations, then by two second-generation stations also named Salyut. All seven were designed to be visited by cosmonauts for relatively short flights. Mir (peace), the third-generation station, was launched in February 1986 as a permanently occupied facility for two or more cosmonauts. Later, three specialty modules—one for science experiments, for example—were added to the basic, forty-three-foot-long main spacecraft. Mir was serviced by unmanned Progress spacecraft, which ferried food, water, fuel, supplies, and mail to it and carried away refuse every three or four months. The station carried scores of Russian cosmonauts and guests from several other countries, including some American astronauts, over a course of twelve years. Huge amounts of data were collected on how humans react to long-duration weightlessness, among many other things. There was also some experimental manufacturing in the weightless environment.

The United States, by comparison, has had a deeply troubled relationship with the space station concept. Skylab, which went up in May 1973, was America's only active station. Its three teams of astronauts used it for a total of only nine months before abandoning it for good in February 1974. It then sailed around the world as an unoccupied derelict for five years before plummeting through the atmosphere as a fireball in July 1979.

NASA quietly studied the concept for a station intermittently at least from the end of Skylab. In July 1983, the space agency, the American Institute of Aeronautics and Astronautics, and several aerospace companies held a three-day meeting in Arlington, Virginia, to work out details for a permanent U.S. station. The spinning-wheel concept gave way to a number of much cheaper alternatives, most of them using tubular modules that sprouted solar arrays, and all of them deployed from shuttles.

In January 1984, President Ronald Reagan announced plans for a permanently occupied station in his State of the Union address. Its original name, Freedom, reflected a Cold War origin. The Canadian Space Agency, the European Space Agency, and the Japanese National Space Development Agency became partners in the program, but they found the United States itself to be highly unreliable. This was because a combination of factors, including budget problems, public ambivalence, technical difficulties, opposition by many scientists, and a lingering feeling that Apollo had made enough of a point about American capabilities, convinced many in Congress that the station was superfluous.

As a result, NASA, the aerospace industry, and Congress went through a long bout of quiet, continuous skirmishes while the station design was reworked again and again. By 1987, a relatively ambitious all-cylinder design gave way to cheaper metal trusses that protruded out of tubular core modules. A graceful, five-hundred-foot-long spacecraft was eventually shrunk down to three hundred and fifty feet. Crew size was reduced from eight to four, and Freedom's experiments were confined to biology and materials processing.

With the end of the Cold War, the name Freedom was quietly dropped, and America's former foes, the Russians, were offered partnership along with fourteen other nations in an all-new design called the International Space Station (ISS).

NASA quietly studied
the concept for a station intermittently at least from the end of Skylab. In July 1983, the space agency, the American Institute of Aeronautics and Astronautics, and several aerospace companies held a three-day meeting in Arlington, Virginia, to work out details for a permanent U.S. station.

While U.S. astronauts lived on the aging Mir for months on end in the mid-1990s, practicing both their Russian and some of the techniques they would need on the International Space Station, the design of the ISS was finalized and work on its core modules was begun.

On November 20, 1998, a Russian Proton booster launched the first space-station module. The forty-two-thousand-pound Zarya (sunrise) was supposed to provide power and propulsion for the early stages of the station's construction. The hard-strapped Russians had built it with American funding. Sixteen days later, six astronauts flying in the shuttle *Endeavour* caught up with Zarya and painstakingly connected it to Unity, the American-built module designed to connect Zarya to other modules.

But the historic connection was only the beginning of a projected five-year construction program that involved the connection of about a hundred additional pieces during eighteen hundred hours of space walks, which would be conducted throughout forty missions. By February 2000, NASA had become so exasperated by Russia's two-year delay in supplying a critically important service module that the agency's administrator, Daniel S. Goldin, threatened to replace it with an American substitute if Russia didn't produce its version within six months. "To say we are frustrated and disappointed is an understatement," he said. Indeed, the entire history of the station was a study in frustration.

Building a space station, piece by piece. The International Space Station's Unity connecting node is hoisted from the shuttle cargo bay before being mated to the Russian Zarya module.

MISSION TO MIR

When the shuttle *Discovery* roared off Pad 39B at 12:22 A.M. (EST) on February 3, 1995, bathing the complex in white light that reflected off billowing clouds of steam, it began a mission of considerable historic importance, both for NASA itself and for international relations.

Mission STS-63 was the first shuttle flight piloted by a woman, Eileen M. Collins, and involved the first space walk by an African American, Bernard A. Harris Jr. And it marked the first time that an orbiter and the Russian space station Mir flew together—a training exercise for a later mission, STS-71, during which an orbiter and Mir would actually dock together.

The high-flying horse race began with *Discovery* trailing Mir by eight thousand miles. By Sunday, February 5, it was less than eleven hundred miles behind the station and closing the distance by ninety miles with each orbit. As rendezvous time neared, flight control teams from both countries worked together to refine plans.

Finally, early on the afternoon of February 6, *Discovery* made its closest approach to Mir, coming to within thirty-seven feet. The two spacecraft flew in tandem, 245 miles over the Pacific, for ten minutes.

"As we are bringing our spaceships closer together, we are bringing our nations closer together," James D. Wetherbee, the mission commander, told the crew of Mir. "The next time we approach, we will shake your hand and together we will lead our world into the next millennium."

Discovery then backed off and gracefully flew around its speeding companion as a final practice maneuver for STS-71.

> Looking at the earth, orbit by orbit, you see no borders. America is not printed with "America." Russia is not printed with "Russia." It is just the planet. It is just my home.

Vladimir Georgievich Titov
mission specialist

Launch of STS-63, February 3, 1995. Onboard were five American astronauts and, for only the second time, a Russian cosmonaut, Vladimir Titov.

MISSION TO MIR

As Bernard Harris Jr. looks on, Michael Foale, who later would live onboard Mir for four months, prepares to grab a Spartan 204 satellite while on the end of the shuttle manipulator arm.

Vladimir Georgievich Titov, mission specialist

A seasoned cosmonaut and veteran commander of Soyuz missions, Titov set a long-duration-stay record in 1988, spending over a year aboard Mir. In 1992, he was one of two candidates selected by the Russian Space Agency for training at Johnson Space Center in Houston.

It was February 1995, six years after I left the station. The Mir had grown larger. When I flew, the station included the base block and the Kvant module and two spacecraft [Soyuz and Progress]. . . . Then we had the approach during our flight STS-63. There were two more modules. It was very wonderful, a wonderful view, and I was very happy because my crew was very happy. . . .

And all of that time we had TV broadcasting from *Discovery* to the ground, and all of the American people could see the station. For me it was a very good opportunity to look at my old home again. At that time, unfortunately, we didn't have a docking unit, so we didn't have a possibility to dock. My dream was, "OK. Maybe I will have one more chance for flight to go aboard the station."

Bernard A. Harris Jr., mission specialist

Titov was certainly invaluable when we got within close approach to the Mir station. As we got closer, the level and volume of the cadence of the Russian being spoken between both spacecraft increased. You could tell they were excited. The Mir crew was just as excited to see us as we were to see them. . . . Here were humans and their machines . . . out there alone in the universe. It was a high moment for all of us.

Janice E. Voss, mission specialist

One of the interesting things about being in a space crew is that there is a whole different language you have to learn, and while Vladimir's English might not have been very strong, his space language was superb. So you could communicate with him on that level. . . . For example, an alarm sounding might read "Secondary Com Failure." He might have not recognized the English, but he could recognize the situation in the space environment. So as time progressed, his English continuously got better. Working with him was a lot of fun because you could rely on him to always do the right thing.

Eileen M. Collins, pilot

STS-63 was Collins's first flight in space.

STS-63 was an extremely challenging flight. We had a little bit of everything: We had the space walks, we had a Russian crew member, we had the Mir rendezvous. . . . We also had a Spartan satellite, on which we did a deploy and retrieve, a robot arm that Janice [Voss] worked, and SPACE-HAB with some twenty experiments in it. It was a challenge for a first-time flier. It was pretty fortunate that I had the opportunity to be exposed to many aspects of the mission because I think that it better prepared me for later jobs supporting the Astronaut Office and for later flights.

Eileen M. Collins

Experience on STS-63 helped prepare Collins for STS-93, during which she served as NASA's first female shuttle commander.

It's usually hard for the first woman in any career field because people are watching you and trying to figure out how a woman is going to do. . . . Because you work harder, you end up doing a better job. For example, in training for my first flight STS-63, there was so much to learn. . . . I really put a major effort into studying. If there were times that I got tired from studying too much. . . . [I'd think], well, I'm not just doing it for me . . . I'm also doing it for other women pilots who want to apply to the astronaut program. I'd like to do well and set an example for them and make it easier for them. That gives you extra motivation. . . .

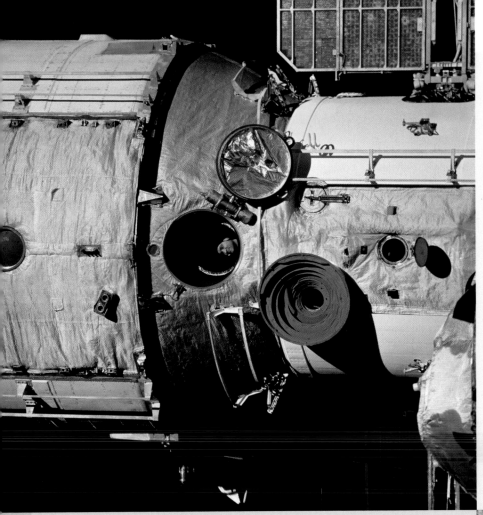

Cosmonaut Valeriy Polyakov peers out of one of Mir's windows at the approaching shuttle. By this time, Polyakov had been onboard Mir for thirteen months.

Below Mir as seen from *Discovery* during STS-63. The shuttle came within 37 feet of the Russian station, verifying that in-orbit rendezvous was possible.

Eileen M. Collins

It was really my third day up there that it really hit me: You're finally here. . . . I was getting myself a drink from the galley, and I thought, hey, stop working for a minute. Just think about this. This is really happening. It surprised me that it took me that long for it to sink in. Part of it was because we were so busy on the flight. But the highlights included seeing my first sunrise in space. Seeing the Mir space station was definitely a highlight. It was beautiful.

Janice E. Voss

I've discovered a wonderful nook up in the corner of the space shuttle by the overhead window where you can curl up next to the window and have a nice place to float. You can stay there easily and read by earthlight. . . . I did that on my first flight with Isaac Asimov's *Foundation*, which is one of my favorite science-fiction stories. . . . My crew gave me a really hard time about it, saying they couldn't believe I was floating right next to the orbiter window and reading a book. (My mother used to say the same thing to me as a child: "I can't believe we're driving through this beautiful Colorado scenery and you're reading a book!") It was such a wonderful experience that I now fly a science-fiction story on every mission. On STS-63, I flew Robert Heinlein's *Stranger in a Strange Land*.

Bernard A. Harris Jr.

Harris worked at NASA's Johnson Space Center as a clinical scientist and flight surgeon until chosen as an astronaut candidate in 1990.

My personal and professional highlight on STS-63 was doing my own experiment right in space that involved the examination of all the crew members and recording their normal findings in space, essentially establishing the foundation for the clinical practice of medicine. It was really rewarding to put in place a program that will allow us, over several missions, to adequately train future physicians. If you can do something to leave a legacy, that's the best thing in the world—the best gratification. Someone can learn from what you have done. You're able to pass it on.

Bernard A. Harris Jr.

During the STS-63 flight, Harris became the first African American to walk in space.

An astronaut dreams of flying in space, but the second dream is to go outside in a space suit. And there are very few people who have done that, a very select group, and I felt very honored to be part of that.

ATLANTIS & MIR: THE LARGEST STATION

O n June 29, 1995—three weeks short of the twentieth anniversary of the Apollo-Soyuz docking—the shuttle *Atlantis* pulled up to the Russian space station Mir and connected with it flawlessly. It was a historic moment because it marked the beginning of a cooperative program in preparation for the building of the International Space Station. It was also the hundredth U.S. human spaceflight.

"We have capture," Navy Captain Robert L. "Hoot" Gibson, *Atlantis*'s commander and veteran astronaut, coolly told Houston as the spacecraft, now locked together, sailed two hundred fifty miles over central Asia. For five days, the shuttle and the Mir that was plugged into a docking collar on its payload bay constituted the largest space station in history, with a total mass of almost one-half million pounds.

Gibson's unemotional announcement soon gave way to a partylike celebration as six Americans and four Russians (two of them carried up on *Atlantis*) shook hands and hugged and kissed one another while chattering in Russian and English. Physician-astronaut Dr. Norman E. Thagard, who had flown a Russian rocket up to Mir and was in his 107th day in space (breaking the American record) said, "You guys are upside down," when he saw the American boarding party.

Anatoly Y. Solovyev and Nikolai M. Budarin boarded Mir, relieving Thagard and cosmonauts Vladimir N. Dezhurov and Gennady M. Strekalov for a trip home on *Atlantis*. Although mission tasks included physiological experimentation, the goal of the joint mission was unabashedly political, not scientific. Its larger purpose was to use space to improve relations on Earth, and STS-71 was only the first of a series of nine Mir-shuttle docking missions that would take place.

▼

The shuttle's given us a very good handle on short-duration space flight, but that is not necessarily something you can carry through directly to a long-duration flight. It's like trying to compare a one-hundred-yard sprint to a marathon.

Andrew S. W. Thomas, flight engineer STS-89

Atlantis returns to Earth, July 7, 1995. Its docking with Mir, apart from being the shuttle's first visit to a space station, signaled a new era of U.S.–Russian cooperation.

The shuttle-Mir missions were an experiment in cultural mixing as much as in technology. Here, the astronauts and cosmonauts involved in the STS-71 crew-transfer mission take a training break at the Systems Integration Facility at Johnson Space Center.

Ellen S. Baker, mission specialist

Baker describes the approach and docking of Atlantis *with Russian space station Mir.*

At first [Mir] looked like a little star in the window. We knew it was the Mir, because it was where it should have been. We all oohed and aahed; it was beautiful. And the closer we got, the more beautiful and exciting it was. During rendezvous we do a few [engine] burns fairly automatically, then once we get closer in, the commander does the flying. Hoot [Robert L.] Gibson was at the aft part of the flight deck, looking out the rear windows and the overhead windows, and using his video camera. We have a monitor inside that he can look at. And Greg [Gregory J.] Harbaugh was using some of our other rendezvous aids, like a handheld laser that gave us distance information. Charlie [Charles J.] Precourt was working our rendezvous software, which gave us more information as far as acceleration, how fast we were approaching, and things like that. All this information was being fed to Hoot and everybody was making sure that it all looked good. Bonnie [J.] Dunbar was talking on the radio to the guys in the Mir, so there was some chatter going on in Russian. The two cosmonauts [Anatoly Y. Solovyev and Nikolai M. Budarin, being carried up to replace the two cosmonauts aboard Mir] were trying to stay out of the way but also trying to get a good view.

Norman E. Thagard, cosmonaut-researcher

Veteran astronaut Thagard was the first American to live aboard Mir, having arrived at the station via a Soyuz launch. His mission tasks included responsibility for twenty-eight biological experiments. In July 1995, after spending 115 days in space, Thagard transferred to the shuttle Atlantis *on flight STS-71 for the trip home.*

For the first few weeks I was fairly busy because we were doing a succession of experiments. . . . However, we started having problems with the freezer. That was the one in which we froze the biological samples, the blood, the urine, that we had taken. . . . So I started spending a lot of time trying to get that freezer to work. Finally it had failed entirely, and then at about six weeks . . . they wound up postponing some of the experiments in the May time frame and I wound up not having enough to do, and that's not good. . . . I mean, there's little by way of entertainment there, and you need to really be busy. . . . My Russian crewmates, from just before the first space walk on the twelfth of May almost to the end of the mission, were chronically overworked. I was chronically underworked. . . . Overwork leads to tension between the crew and mission control, is what I saw. So you don't want to be at either extreme; you want to be somewhere in the middle.

Norman E. Thagard

Thagard describes the docking of Atlantis *with Mir, after several launch delays.*

It's a pretty awesome sight. I had really thought that with two two-hundred-thousand-pound vehicles coming together in orbit that everybody's eyes on both Mir and the shuttle would be about yea big and everybody would be nervous, but when it actually came to the event that wasn't true at all except for maybe Hoot [Robert L. Gibson], because he had to fly the shuttle. It looked like it was so slow and so well controlled that you just never felt like anything was going to get out of hand. In fact, it progressed very nicely and very smoothly into a dock, and it was very precise, I thought, very well done. . . . I remember over air-to-ground I said to Hoot as he was coming up, "Well isn't that the way it is? You call for a taxi, and it takes weeks to get here."

Ellen S. Baker

It was a thrill for me to be on Mir. First of all, the living space is just giant compared to what we're used to in the shuttle. In fact, it's so big you can almost get a little disoriented going from module to module—it's not always laid out in a way that an Earth-based person would think intuitive. Sometimes the modules are rotated in relation to one another, and you come into the node where all the modules meet, and some of them will be rotated, and not all in the same orientation. You have to ask yourself, "Where am I? Which way do I really want to go here?"

Shannon W. Lucid, board engineer

Lucid was the second U.S. astronaut to fly aboard Mir, boarding the space station in March 1996 from STS-76, and remaining on it for six months.

The one big take-home message from Mir that I've been trying to get across is that the most important system that will be there on space station and needs to work really well is the crew dynamics and the crew interactions. Every day I realized how fortunate I was getting to work with Yuri [Onufriyenko] and Yuri [Usachev]. . . . We did gather together for our meals and this made a big difference in your daily life and in your contentment in working together. I know some crews did not do that and I think that they suffer because of that. It's just like a family. I always thought it was very important as a family that we always have some time every single day where we gather together to eat. It's very important that the facilities on a station allow for this to happen.

Andrew S. W. Thomas, flight engineer

Thomas boarded Mir in January 1998 from the shuttle Endeavour *as part of the STS-89 crew and served as flight engineer there for 130 days. He returned home on STS-91,* Discovery, *on June 12, 1998.*

You have to find creative ways of getting recreation and keeping yourself psychologically refreshed in an environment that has very few options available to you. Once I'd overcome that and settled into a routine, then things progressed very easily and it became actually a very enjoyable experience because I was then in a position to enjoy the uniqueness of the environment that I was in. . . . As I look back on it, I think in the closing part of the twentieth century in spaceflight, it's probably one of the most unique experiences that you could possibly have.

C. Michael Foale, board engineer

In May 1997 Foale boarded Mir from Atlantis *on STS-84, spending 134 days there, during which he took part in an EVA to inspect damage caused by a collision with the* Progress *resupply ship. He returned with the crew of STS-86 aboard* Atlantis *in October 1997.*

I was very much aware of a social or cultural difference between our Western society—and it's really a Western thing and not just an American thing—and Russian society and cultural expectations. . . . I wanted to be as tough as the Russians in handling difficulties and hardships. And I think I did that well and the Russians recognized it, and toward the end of my flight really made an effort to let me do a space walk in their space suit, which was never planned. I was qualified to do it, but it was not necessary. It was not the most logical thing for the Russians to have done, to put me in a space suit rather than the new Russian coming up in the Soyuz. They did this, I think, to reward me for my stoicism.

THE SPACE STATION COMES TOGETHER

A new breed of construction worker was born on December 6, 1998: the orbiting hard hat.

That day, astronauts aboard the orbiter *Endeavour* joined the first two segments of the International Space Station two hundred and forty miles in the sky. After spending most of a day chasing the forty-thousand-pound Russian Zarya control module and finally snaring it, the spacefarers carefully positioned it over the shuttle's payload bay, aligned it with the American-made Unity docking port, and then delicately raised *Endeavour* until both sections could be joined and clamped tightly together.

Unity and Zarya were the first of one hundred major components that were scheduled to be connected in roughly eighteen hundred hours of extravehicular activity (EVA) on some forty missions extending over a five-year period. Noting that the chronically ailing Russian economy was two years late in delivering a service module that included living quarters, though largely paid for by the United States, skeptical observers predicted that completion of the station would be long delayed.

The new hard hats were well aware that they worked in an exceptionally high-risk environment, often wrestling with huge components, methodically and precisely using tools to connect them in an airless, bitterly cold, gravity-free void while moving at more than seventeen thousand miles an hour.

Four days after Zarya and Unity were joined, their first occupants came onboard. Shuttle commander Colonel Robert D. Cabana and Russian cosmonaut Sergei K. Krikalev first climbed into Unity and then into Zarya. On December 13, they set the embryonic station loose and returned to Earth.

The Russian module
Zarya—the functional cargo block of the new International Space Station—launches from Kazahkstan, Russia, on November 20, 1998.

Orbiting construction workers Jerry Ross (left) and James Newman make their final connections to the first pieces of the station during the last of three space walks conducted in December 1998.

ENDEAVOUR

James H. Newman, mission specialist

During the mission to mate the Unity and Zarya modules, Newman, along with Jerry L. Ross, performed three demanding space walks, totaling over twenty-one hours, to connect power and data transmission cables.

STS-88 was more momentous than my previous flights in the sense that I felt very, very privileged and lucky to be on the flight that kicked off what we hope will be a new era in our space program—an era of international cooperation on an unprecedented scale, and a permanent space station. . . . It was very exciting to be up there with the first modules. The space station is really a marvelous place, even at the size it is now, and it's only going to get bigger and more special.

The joined Zarya (left)
and Unity modules are
surveyed by the STS-88
crew in a fly-around
hardware inspection.
The modules are the first
of more than one
hundred pieces to be
mated in space during
the next few years.

Sergei K. Krikalev, mission specialist

Krikalev helped assemble the first two pieces of the International Space Station, where he will live as part of the first three-man crew.

I remember when I flew for the first time on Mir, and my experienced crewmates told me how things would go in space and what to expect on a station. They were things I learned from my friends. So on STS-88 I tried to be helpful, to talk to the new guys and explain to them what they should expect. . . . Working inside the station, I could explain how I would do something. Sometimes it's very difficult being in weightlessness. For example, to stay stable and point a camera accurately and take a picture properly, there are many things you have to keep in mind—not only film speed, aperture, and shutter speed, but also how to attach yourself, how to find what you really need to make the picture. So I think my previous experience of many months in space helped.

Gregory J. Harbaugh, EVA program manager

Constructing the International Space Station will require an unprecedented number of rigorous extravehicular activities (EVAs) in space. Harbaugh will direct the space-walking crews who will do much of the work.

The EVA assembly time required for the station has been around 160 to 165 EVAs, which has been pretty consistent for about two and a half to three years. . . . We have a chart that shows the buildup of the time over the next five years. It shows what we did starting back in Gemini and then Apollo, walking on the Moon. Those are little bumps down at the bottom, and then you get up to space station assembly and there's this huge spike over the next five years. It is a big concern and is something that requires constant vigilance. The good news is that if you break that total package down into discrete elements, what we plan to accomplish on any given shuttle mission and the station-based [EVA] work, there's nothing there that's outside of our experience base. In other words, there are only a couple of five-EVA missions and we've done two highly successful Hubble [Space Telescope] missions that were five EVAs, so we know what we can do on a per-shuttle basis.

The big concern, the big challenge is to ensure that when we get there, all the pieces fit, that we really understand the interfaces, the access to the work sites, the body positioning, those kinds of things. Something as monstrously big as the space station requires an awful lot of effort to verify that you can access every place where you might have do work via EVA, and that you can do what needs to be done. . . . We have planned that for every hour of EVA done in orbit, we'll do ten hours of training in the water. That template has been borne out over the last several years as a valid ratio. . . . We've supplemented that training time in the water now with time in the virtual reality laboratory, which can provide crew members the opportunity to step through the process of executing the EVA. It helps much more than anything we've had previously.

Randy H. Brinkley, program manager

The International Space Station was born out of national competition—competition between the East and the West. It was a way that, just like going to the Moon, we were going to demonstrate the unity of the Western democracies among spacefaring nations, as compared to the Russians. It has subsequently evolved from national competition into international cooperation by pulling the Russians into the partnership, which I think is a very important thing. I think it will be one of the things that, ultimately, will be very important in terms of its success, as difficult as it has been.

Franklin Chang-Diaz, astronaut

Prior to his flying on STS-91, the final shuttle-Mir docking mission that concluded the joint U.S.–Russian Phase I Program, Chang-Diaz participated in the early design phases of the space station.

We have been on a very long road in evolution and growth of the space station, and I think that has been necessary for us to be able to embark on a long-lasting effort. . . . Slowly people are beginning to focus on the reality of having a permanent outpost in space that will allow humans to live and work there in a multinational sense. I think the future is very clearly a multinational effort to go into space.

Space shuttle

Endeavour, with the Unity module (aka Node 1 of the ISS) in its cargo bay, approaches the Zarya module, which has been waiting in orbit for just over two weeks.

Kathryn "Kathy" Clark, chief scientist

What the station is really going to do is provide a laboratory in an environment that cannot be duplicated anywhere else. Most of the time when you get a huge breakthrough in medicine, in any other research field—material processing or combustion—it is because you have been able to look at something from a unique vantage point. For instance, when the electron microscope was invented, all of a sudden many discoveries were made. It was because we could do something that we could not do before. We could see things that we could not see before. The station provides that big of a difference because the environment is so completely counter to anything that we have had on the earth. We have had a taste of it with the shuttle but with very short access.

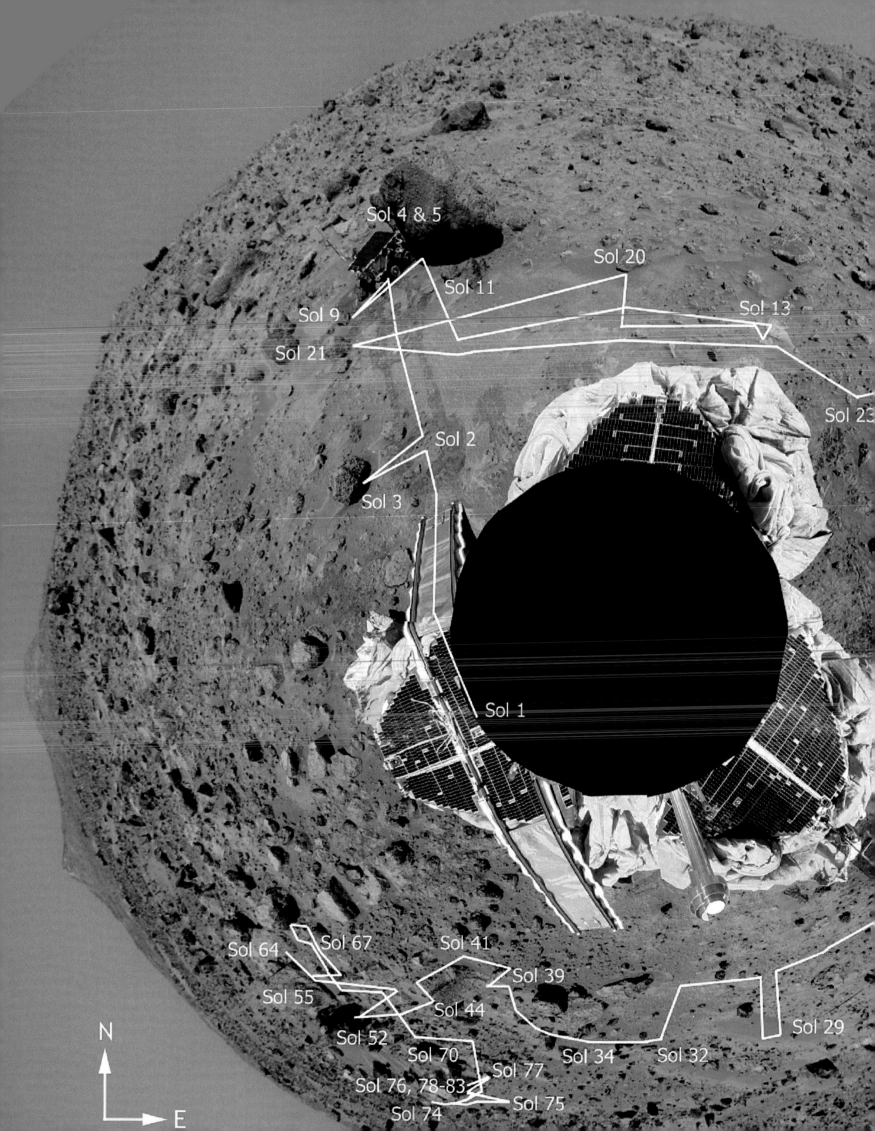

RETURN TO THE
RED PLANET

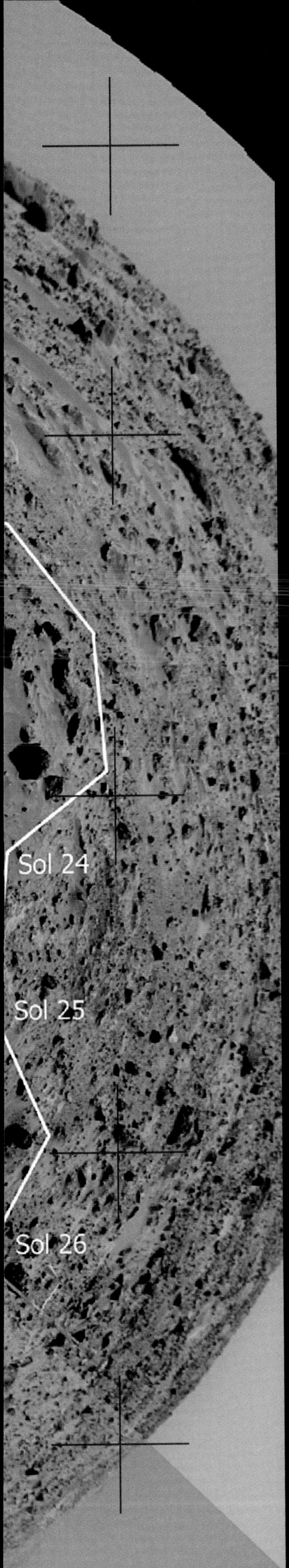

F or scientists and engineers like Wernher von Braun, Sergei
P. Korolyov, and uncounted thousands of their colleagues
and disciples, not to mention science-fiction writers from H.
G. Wells to Isaac Asimov, the Moon and space stations were
merely jumping-off places on the way to the always alluring and enigmatic grand prize: Mars.

The Viking missions of 1976, with their paired landers and
orbiters, sent back a torrent of new data on the planet's varied topography (which suggested that at one time there were deepwater oceans,
possibly containing life), its atmosphere, winds, weather patterns, iron-
oxide-rich soil, a landslide in progress, geological activity at Olympus
Mons, Valles Marineris, and elsewhere, and a great deal more.

Yet two factors captivated the space-science community and
the legion of Mars devotees. The mass of data inevitably (and correctly) raised new questions that only whetted their intellectual
appetites, and the results of the three life experiments conducted by
Viking while on the Martian surface were so ambiguous that they
begged for a closer look. A return to Mars, however, would prove
both exhilarating and frustrating.

In 1983, the NASA Advisory Council's Solar System Exploration Committee recommended a series of low-cost, modestly scaled

Overhead view of the
Mars Pathfinder land-
ing site, showing the
path traveled by the
Sojourner rover during
the summer of 1997.
The camera-equipped
robot spent eighty-
three Martian days
(one day equals about
24 hours, 37 minutes
Earth time) exploring
the surface, examining
rocks, and analyzing
their chemistry.

Sol 24

Sol 25

Sol 26

missions to the inner planets that were collectively called Planetary Observers. The first in the series, Mars Observer, was designed to swing into a low polar orbit over the Red Planet and repeatedly map its surface and atmosphere, defining topography and the gravitational field, in great detail every 59 days for the 687-day Martian year. But Mars Observer was plagued by cost overruns and development problems on Earth and, ultimately, by the "Great Galactic Ghoul" after it left home. The ghoul, a fearsome figment of the imagination, was invented in the 1960s to explain why Mariner 7 and other Mars-bound spacecraft, several of them Soviet, developed otherwise unexplainable maladies and either went haywire or crashed.

The fifty-seven-hundred-pound Mars Observer left Cape Canaveral in apparent good health on September 25, 1992. But much to the horror of its stunned team, the first spacecraft to visit Mars in seventeen years abruptly went silent the following August, just as it arrived. A postmortem the following year revealed serious design flaws and operating procedures by the Jet Propulsion Laboratory (JPL), which managed the $980-million spacecraft.

But deep-space exploration, always a risky enterprise, took a happy turn four years later. On the Fourth of July, 1997—twenty-one years after the Viking landings—a spacecraft named Mars Pathfinder dropped out of the ocher Martian sky and bounded in inflated bags safely across the rocky plain called Ares Vallis. On July 6, with members of its team cheering at JPL, Mars Pathfinder's petal-like solar panels opened to reveal a twenty-five-pound, six-wheel rover named Sojourner. The little robot obediently rolled down a ramp and instantly made history by becoming the first machine from Earth to tour the surface of Mars. Sojourner spent days chemically analyzing several rocks it bumped into, three of which were quickly named Yogi, Barnacle Bill, and Scooby Doo. Pictures of rock beds taken by Mars Pathfinder showed that they were tilted in a way that suggested the movement of a great deal of water. Perhaps more important, the spacecraft sent back the first unmistakable evidence that Mars, like Earth, had a crust, a mantle, and an iron core. This, too, suggested that it might indeed have once held life. Mars Pathfinder was eventually renamed the Carl Sagan Memorial Station in honor of the enthusiastic astronomer.

Mars Global Surveyor (MGS), another relatively cheap spacecraft, was an international explorer intended to be the first in a long series of "faster, better, cheaper" missions to Mars: robotic missions that would effectively establish a permanent planetary presence. Arriving two months after Pathfinder, it used a pair of instruments to map the planet in three dimensions in exquisite detail, one being the Mars Orbiter Laser Altimeter, which fired infrared laser pulses at the terrain and measured their return time. MGS was so successful that the journal *Science* carried six reports from its science teams and ran a startlingly clear photograph of part of the Valles Marineris canyon on the cover of one special issue and two composite color photographs of the planet on another cover.

By sending back that imagery, Mars Global Surveyor demystified a longtime staple of the supermarket tabloids: the supposed human face that seemed to appear in a 1976 Viking orbiter picture taken of steep terrain. An MGS picture of the same place revealed the eerie site as being merely a bumpy mesa that, if anything, resembled a sandal print.

▼

The little robot obediently rolled down a ramp and instantly made history by becoming the first machine from Earth to tour the surface of Mars.

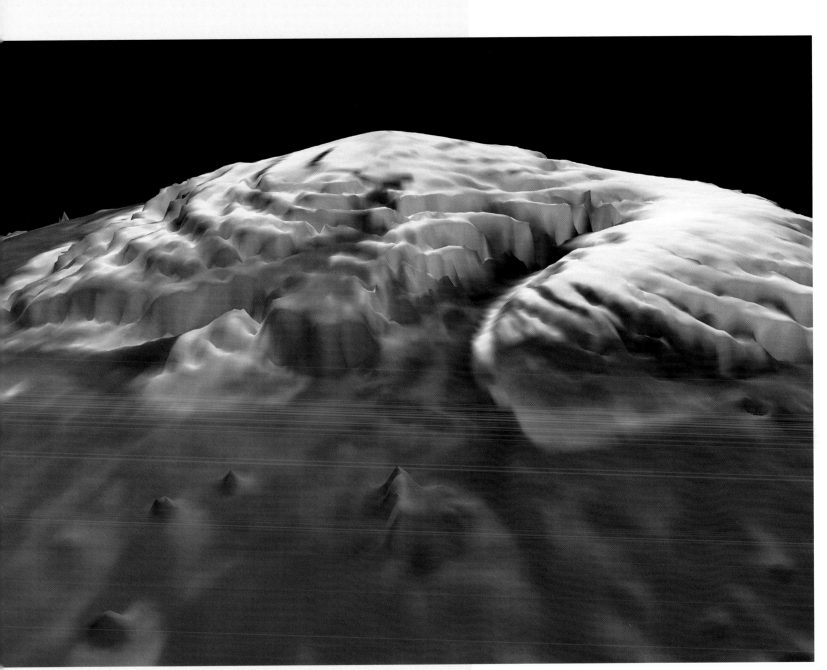

The winning streak with Mars abruptly ended in the autumn of 1999 when two other faster-better-cheaper probes, Mars Climate Orbiter and Mars Polar Lander, ran into catastrophic trouble. The orbiter apparently plowed into the Martian surface when it came in at a steeper angle than was expected because its navigators made trajectory errors and failed to realize that it was dangerously off course until it was too late. In early December 1999, the lander, designed to use a mechanical arm to search for water-ice at the south pole, simply disappeared for no apparent reason.

No one believed that the setbacks would break the growing bond between Earth and Mars, however. The script still called for orbiters and landers to be dispatched to Mars during each two-year window through 2010. And starting in 2003, new rovers—larger descendants of Sojourner—were supposed to scoop up soil samples that would later be tossed into orbit for relay back to Earth by rocket. Few doubted that there would occasionally be more trouble during the retrieval missions. Fewer still doubted that the trouble would be worth it.

The Martian north pole with its cap of water ice appears in 3-D in this computer-generated view produced by the Mars Global Surveyor. A laser onboard the spacecraft pulses the ground, measuring the height of surface features.

MARS PATHFINDER

Mars Pathfinder, which got a good bounce in both senses of the term when it landed in Ares Vallis on the Fourth of July, 1997, represented a new breed of Mars explorer.

Mars Pathfinder looked at all of the rocks around it and sent back unmistakable evidence that long ago the planet had a watery deluge of immense proportion. The one-time presence of water was not a surprise; Ares Vallis was picked as a landing site precisely because it was an ancient floodplain. But the observation that it appeared as predicted from orbital remote sensing at kilometer scale, was surprising indeed.

Dr. Michael Malin, a geologist and noted space-camera designer, analyzed the shapes of the rocks, their colors and overall position, and the shapes of ridges in the area. He concluded that Ares Vallis was the scene of a fast-moving flood that was hundreds of miles across and hundreds of feet deep, and that flowed for thousands of miles.

No one needed to remind Pathfinder's team of young scientists and engineers—many of whom were in grade school when Viking 1 landed 527 miles from their spacecraft in 1976—that every place on Earth that has water has life.

Sojourner, Pathfinder's twenty-five-pound rolling robot, did the first chemical analyses of Martian rock. They showed that the planet had a dramatic geological history marked by repeated cycles of internal melting, cooling, and remelting. Excited mission scientists called the findings a "rock festival."

▼

Pathfinder had no national political motivation. It was the equivalent of a bunch of guys doing something in their garage because they loved to do it.

Matt P. Golombek, chief project scientist

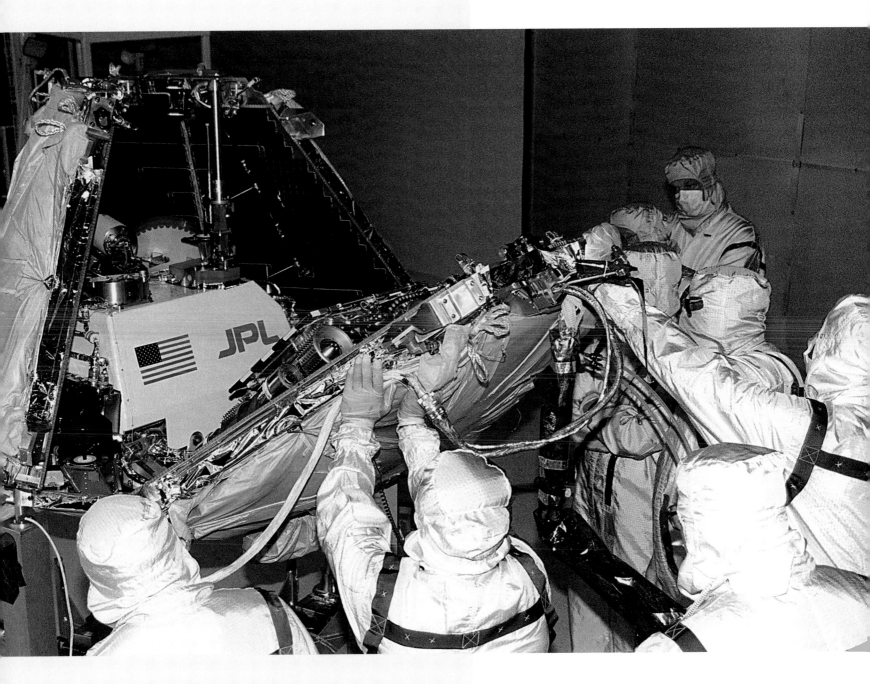

Preparing Mars Pathfinder for launch.

Tucked inside the lander are the Sojourner, the
Imager for Mars Pathfinder (IMP), and other instrumentation and equipment. The Pathfinder was the
first spacecraft to visit Mars since the Vikings of
the 1970s, and was built for a fraction of their cost.

MARS PATHFINDER

Donna Shirley, manager, Mars Exploration program

The biggest challenge for me was getting the rover a ride to Mars. The scientists had wanted a pickup-sized truck rover . . . and we were developing the technology for it. But when the budget was cut at the end of the Cold War it became really clear that we couldn't spend $6 billion on a sample return mission, so it had to be downsized.

Dave Miller [technical group supervisor] had a group working on small robots with technology based on insect behavior, technology developed at MIT. The robots are not very smart but smart enough to pick up a sample and bring it back to the lander. And so we worked on a small rover that could accomplish some of the objectives of a big rover. Then the Germans came along with an Alpha Proton X-Ray Spectrometer [APXS] that was small enough that the rover could actually carry it. So we sort of bartered: We'll take the instrument along and put it on a Martian rock to see what the rock is made out of if you guys will take us to Mars.

Above

Halfway through its exploration of Mars, Sojourner travels between two rocks nicknamed Wedge (left) and Flute Top (right). The cylinder protruding from the rover's back is a spectrometer used for analyzing rock chemistry.

Brian K. Muirhead, project manager

The bottom line for success in doing things faster, better, cheaper on Pathfinder was the team; we would put people before process, before rules. We took license to invent how we would do this job. It was such a daunting set of constraints: the time, the budget. . . . If you look at what we were trying to do—$150 million in 1992 dollars compared to Viking at $3 billion in equivalent dollars—it looked impossible. In fact, a lot of people didn't think we'd even get to the launchpad, let alone the surface of Mars. So it was clear to JPL and NASA management that we were going to have to be free to take from the past what we thought would work, and then pretty much invent everything else. And to be sure that all of our new ideas would work, we tested the hell out of them. "Test. Test. Test."

Matt P. Golombek, chief project scientist

To select the Pathfinder landing, we used the highest resolution images of Mars that we had, which was thirty-eight meters [one hundred twenty-five feet] per pixel. This allowed us to identify things in an area about the size of a football field. Of primary interest to landing safely are objects that are about the size of your desk. We were landing blind.

My greatest success was that we accurately predicted a rocky landing site from those images. The goal of Pathfinder was to land in a place with a variety of rocks so we could interpret the history of the planet from them. When the first images came back, I was jumping up and down, yelling, "Rocks! Rocks!"

This 360-degree, color-enhanced panorama of the Pathfinder site was taken over the course of three Martian days. The unfolded "petals" of the lander lie on top of its deflated airbag and the forward and rear rover ramps are visible. The rock Barnacle Bill stands at near center.

Cameras on the rover show fine texture in a boulder nicknamed Chimp. The sweeping "tail" of sand beginning at lower right was most likely caused by the Martian wind.

M A R S   P A T H F I N D E R

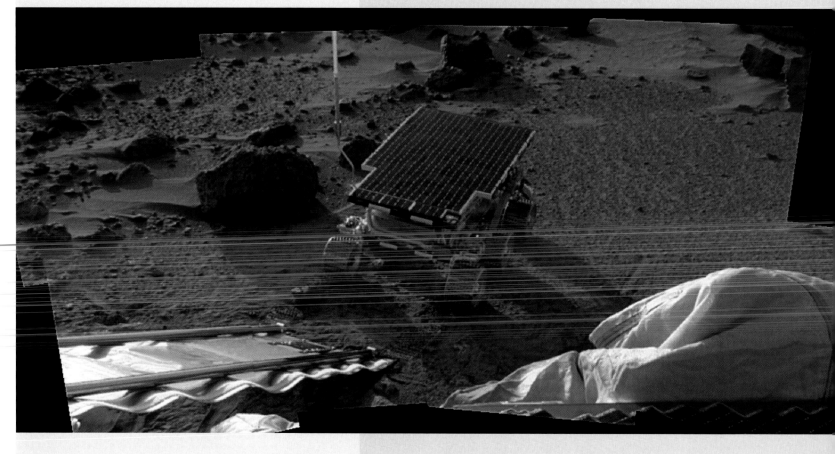

Matt P. Golombek, chief project scientist

We learned a tremendous amount from Pathfinder. We found a certain kind of rock that could have been formed in running water, and we found an abundance of sand that typically forms in water or due to the action of water. We found out that the dust in Mars's atmosphere is extremely magnetic. Interpretations for the origin of that magnetic dust are very suggestive that it formed in a water-rich environment. All of this suggests that Mars might have been a warmer, wetter place than it is now. Right now the pressure and temperature on Mars make liquid water impossible. But if the environment on Mars was similar to the environment on Earth when life started here, about 3.6 or 3.9 billion years ago, we want to know if life formed on Mars, too. If it didn't, why not? And if it did, what happened to it?

Above

Late afternoon on Mars. Sojourner casts long shadows in the waning sunlight, not far from the ramp down which it rolled to reach the surface.

John D. Rummel, NASA planetary protection officer

Rummel is responsible for ensuring that NASA spacecraft do not biologically contaminate solar-system bodies, including the Earth. The strict protocols followed for the Viking expedition to the Martian surface in 1976 yielded information that contributed to the planning of Pathfinder.

The Viking craft itself, the lander, was actually baked in an oven. It spent about fifty hours in the oven and at over 110 degrees Celsius (230 degrees Fahrenheit), just to kill off Earth contamination. That was for two reasons: to protect Mars and to protect the science results from the life-detection experiments.

Now we understand enough about Mars to believe that Earth organisms won't spread, so we're able to do missions in a much less demanding engineering environment. The Pathfinder mission is one example of a mission that could not have been flown under the Viking planetary protection requirements. We don't think that any of the very large integrated chips that were in Pathfinder's computers could be baked in an oven and survive. I don't think that the airbag system itself could have stood that kind of heating, although the parachute was certainly heat-treated. That doesn't mean that Pathfinder was dirty spacecraft. Basically it was very clean, with fewer live microorganisms on it than you'd probably find in a glass of bottled water. We still clean them up, we just don't heat-treat them.

▼

The most rewarding part of the mission for me was seeing that rover crawl down the ramp and put six wheels on soil. That was a real peak experience. That was absolutely the best.

Donna Shirley, manager, Mars Exploration program

This 200-degree panorama shows the Twin Peaks in the center distance. The hills helped scientists determine Pathfinder's location from orbital photos. North is to the right.

MARS GLOBAL SURVEYOR

A new era in the exploration of Mars began on November 7, 1996, when Mars Global Surveyor (MGS) left Earth on the first in a series of international missions designed to build on Mariner and Viking data and establish a permanent robotic presence at the Red Planet.

"Mars Global Surveyor does not just represent a single launch. It opens a whole new era to do planetary exploration both systematically and economically," said Wesley T. Huntress Jr., NASA's associate administrator for space science at the time. The more than one-ton spacecraft, meant to replace the lost Mars Observer, was built with many of the Observer's leftover spare parts.

MGS arrived in September 1997 and spent the next four months gradually lining up into a nearly circular, two-hour, south-north polar orbit that averaged 234 miles above the Martian surface. It then used its specialized instruments to collect the most detailed information ever on the Red Planet.

The spacecraft's most dramatic contribution was the first three-dimensional map of the planet made with a laser altimeter that took twenty-seven million contour readings. Scientists were amazed to see that the Hellas basin, which is thirteen hundred miles across and six miles deep, is really an ancient impact crater that would stretch from Maine to Colorado.

Mars Global Surveyor found evidence of an ancient, cataclysmic flood that probably contained enough water to fill a one-hundred-yard-deep basin the size of Utah. It also discovered magnetized bands in the Martian crust that indicated the planet surged with internal heat and other powerful forces relatively early in its life, just like Earth, before both planets went their own very separate ways.

▼

The laser altimeters on Mars Global Surveyor have an incredible precision of thirty centimeters [about a foot]. By the end of the mission, we'll have enough measurements to make a map of Mars that would be comparable in detail to a map of small back roads in a rural area.

Timothy J. Parker, geologist,
Jet Propulsion Laboratory

Close-up of Martian canyon Coprates Catena, part of Valles Marineris. The Mars Orbiter Camera on the MGS spacecraft has an unmatched ability to show fine detail—revealing features on the surface as small as 3 feet across.

SURVEYOR

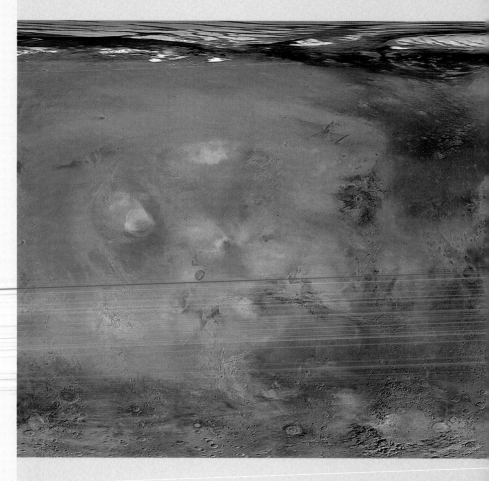

This Martian weather map, taken on a single northern summer day in April 1999, was assembled from twenty-four images snapped by the MGS wide-angle camera. Wispy bluish white clouds hang over volcanoes in the planet's Tharsis region.

Michael C. Malin, principal investigator, Mars orbiter camera (MOC)

Malin Space Science Systems, of which Malin is president, built and operates the MOC, which obtains images of the surface and atmosphere on Mars for study in the areas of geoscience, meteorology, and climatology.

Imaging Mars is a continuous process. Twice a week, Ken Edgett, who has really targeted 99 percent of the images, or I come in at six o'clock at the morning. And we don't leave until six or seven in the evening, and we target the next three or four days' worth of images. We can't get further ahead than that because the orbit prediction isn't good enough to tell where the spacecraft's going to be. So you have to use the very last update you get to figure out where we're going to be and what we can take pictures of. And while you're targeting for the next four days, images that you targeted half a week earlier are already coming down. We get a couple of hundred images a day sometimes. There are only a few people—and I'm not one of them—who've seen every picture that's been taken.

Timothy J. Parker, geologist, Jet Propulsion Laboratory

The existence of a primordial ocean on Mars has been a topic hotly debated by the scientific community. Parker is one of the strongest proponents of the hypothesis that a long-standing body of water once occupied northern Mars.

Using Viking image data, I had looked for clues to what happened to the gigantic floods that created the big channels on Mars, particularly in the Chryse region. They dwarf even the biggest flood channels on Earth. The capacity of the channel dimensions imply something on the order of ten thousand times the output of the Amazon River. What happened to the water? I started in the Chryse basin and worked my way to either side. I was surprised at how far I had to go before I found anything that looked like a shoreline. I traced what I thought were two shorelines, one inside the other, all the way around the northern plains—25 percent of the planet's surface area. The only other place in the solar system that is as flat as this area of Mars is the earth's seafloor. The shoreline I found indicated that this was once a liquid ocean.

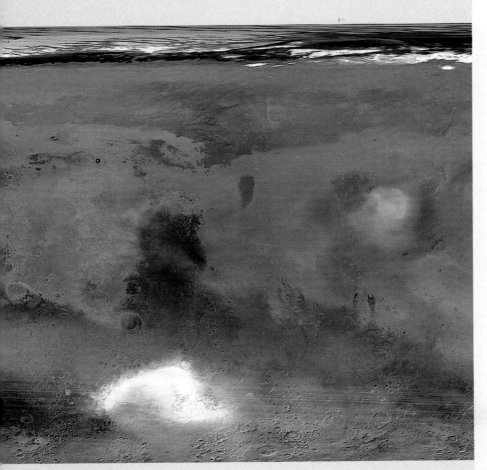

Michael C. Malin

We've gotten a bunch of pictures where it's been easy to put myself there mentally—some pictures of sand dunes, in particular, because the dunes on Mars and the dunes on Earth are so remarkably alike. Because of that and because the scale that we're looking at is just the scale of aerial photography, and also since I spent a lot of time walking on dunes in Antarctica and in Iceland, it's easy to put myself into these pictures and imagine how long it would take me to climb up [a Martian] dune and move along its crest. I can almost vicariously feel what it would be like to slide down the slip-face of the dune and hike up over to the next one because I've done that enough times to know. As soon as I start looking at a picture, I start imaging what it would be like on the ground.

But I should point out that there are many, many places on Mars that I haven't the *slightest* idea what they would look like on the ground. I'm still mystified by the planet. And in fact, we're seeing many things where we believe we will not actually be able to tell how they formed until a human is out there walking around on the ground. It has become very profound—almost a kind of melancholy—for my colleague Ken Edgett and I, because we know that we're not going to get to do this. All the things we are now seeing that we're very excited about, we know that none of the currently planned missions are going to address. It's very dispiriting in a sense. It also, obviously, fires our imaginations. Neither of us sleeps much at night.

I talked about this with colleagues over lunch. They smiled mostly. At that time, scientists thought that the catastrophic floods froze or soaked into the ground very quickly. They wouldn't accept that the floods could have collected into a standing body of liquid for any length of time.

I made the first public presentation of this idea in Washington, D.C., in 1986 and it was pretty skeptically received. It took until 1989 to get a paper published about it. My reticence at getting that first paper out had to do with anxiety. I was really worried that I would be laughed at, and I think I was by some people.

What convinced many of the remaining skeptics was the high-resolution topography data obtained by Mars Global Surveyor. I was absolutely goose-bumpy when the data showed the interior of the two shorelines to be approximately level all the way around. Those who were on the fence were pushed over to the "wet side." Now, MGS is finding evidence of crystalline hematite in what would have been a submarine environment, if my shorelines are real. That gets me all excited. Crystalline hematite had to have been formed in a liquid water environment.

▼

Mars is a small planet that
does things in a very big way.

John Pearl, Goddard Space Flight Center

SURVEYOR

Phil Christensen, principal investigator, thermal emission spectrometer.

The TES measures the thermal infrared energy, or heat, emanating from Mars. The information gathered by this instrument provides data for study of the planet's clouds, atmospheric dust, rocks, soil, and ice.

I have always been curious about what Mars was like millions of years ago. Since we can't go back in time I have tried to work on ways to read the history of Mars from the geological record locked within the rocks on its surface. One way is to look at the minerals present. If, for example, Mars did have large lakes or hot springs, then certain types of minerals would have formed and may still exist today to provide us a window into the planet's past.

For over fifteen years I have had the pleasure of working with a wonderful group of engineers to build and fly experiments to Mars to try to unravel the story in these rocks. Our second instrument, called the TES, is currently in orbit with Mars Global Surveyor (the first is somewhere orbiting the Sun on Mars Observer). Early in the mission we struck "gold," so to speak (actually, it was a mineral called hematite). TES had detected hematite, which forms in water. The few places where it occurs are even more interesting because of its extreme rarity, but it always occurs in places that suggest other signs of liquid water. With this single discovery I felt that we had found what we had come looking for—direct evidence for water—and fifteen years of effort on the TES project had been justified.

However, for me, scientific discovery is just the final icing on the cake. The real pleasure in exploring space is the process itself: developing an idea—literally on scraps of paper with friends over dinner; shaping the idea into a concrete proposal; working through the inevitable setbacks and problems that occur when you first build something; figuring out ways to get enough money to finish the project; motivating people to work a little longer; finally seeing the idea take shape as an actual instrument; and sometimes—as was the case with me with Mars Observer—watching a mission fail then starting the whole process over again.

One of my most enduring memories is standing at the edge of a lab crammed with a team of six or seven incredibly talented engineers and all of their test equipment as they huddled around the TES trying to understand a serious problem. It looked like an operating room with TES at the center. For hours the room was silent as each person worked and analyzed their piece of this complex puzzle. Every now and then the silence was broken as someone would pose a question and suggest a test, "what if we try . . . ?" Outsiders would walk in and fall silent—immediately recognizing the level of concentration that was occurring in that room. Finally, after three long days of work, frustration, and intensely focused energy, one of them said very simply and quietly, "I know what it is."

> We get calls from other scientists, and they come here to look at the pictures. You see them wandering up and down the hallways, shaking their heads. And they'll come and stand in my doorway, and say, "What's going on in these pictures? I don't understand!" They're just baffled. Mars is such an amazing place, and our lack of understanding is so appallingly huge.

Michael C. Malin, principal investigator, Mars orbiter camera

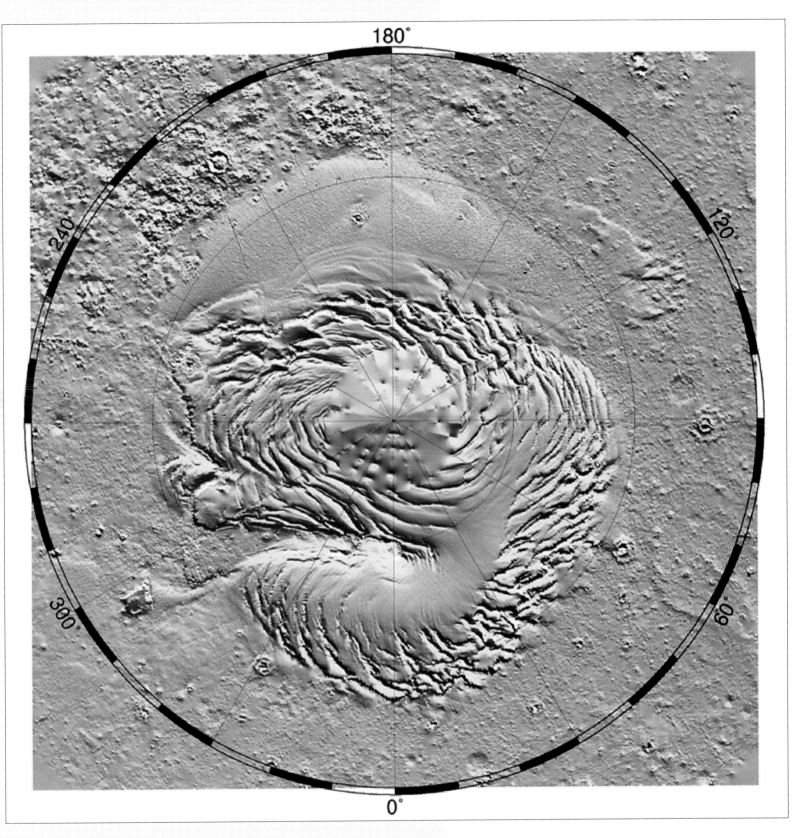

A relief model of the topography of the north

polar region of Mars depicts the ice cap and its

surroundings. The Surveyor's Mars Orbiter Laser

Altimeter (MOLA) instrument measures surface

topography with great precision.

THE END OF
THE BEGINNING

Within the span of a single century, airplanes have gone from rickety wooden contraptions covered with stitched-on fabric to jet-propelled airliners that routinely carry four hundred passengers across the Atlantic at speeds and altitudes that would have astounded the Wright brothers. If space travel develops at the same pace—and no one in either aviation or space seems to doubt it—significant numbers of people are destined to orbit Earth and go from there to the Moon, Mars, and beyond.

Highly informed dreamers like Buzz Aldrin, others in the astronaut corps, and everyone in the vast scientific and industrial infrastructure who stands behind them have no doubt that it is humankind's destiny to migrate to other worlds: to extend a human presence beyond the home planet for adventure, commerce, and survival. Aldrin, who holds a doctorate in astronautics from the Massachusetts Institute of Technology, has stood on the Moon, and is a very realistic businessman who dreams. But he does not dream idly. As these words were being written, in fact, the just-turned-seventy entrepreneur had engineers (not animators) designing a reusable booster system that could carry a hundred paying customers to low Earth orbit for quick round-the-world jaunts or for longer stays in orbiting hotels. Rather than dropping into the ocean and sinking like rocks, these winged robot boosters would swoop back down to Earth for refurbishing and then fly again. And again. The Starboosters, as he calls them, are not only technologically feasible but potentially profitable.

Others are just as convinced as Aldrin that tourism in space will be—not could be—a logical extension of the unprecedented boom in huge ocean liners that increasingly cross this planet's seas carrying tourists. Plans are in the works to develop other reusable spacecraft. Some of them would use a single-stage booster to orbit, launched straight up like today's rockets. Others would be pulled up by helicopter-like rotor blades supplementing a rocket, or would be composed of an airplane-like rocket liner towed by an airplane high enough to give it a running start to orbit. One Boeing design features an upgraded shuttle with "flyback" boosters. As with airliners, reusing space liners will make them both profitable and safe, since their engines will be repaired, refurbished, or replaced.

In December 1999,
Hubble Heritage
Project team members
acquired data from the
Hubble Space
Telescope to arrive at
this image of NGC
1999, a nebula found in
the constellation Orion.

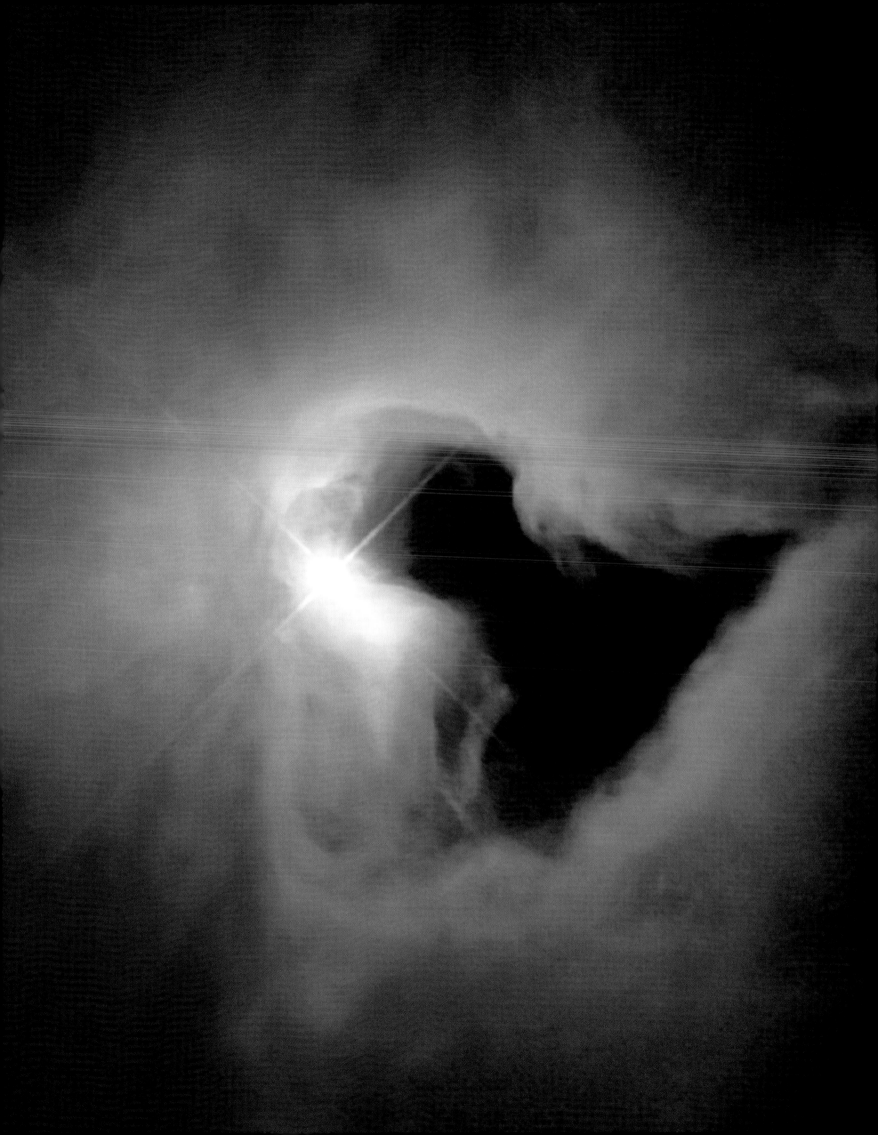

The Artemis Society International, a privately funded, self-described "lunar underground" whose members are engineers dedicated to colonizing the Moon, plans to create a mile-square "lunar base" near Las Vegas, Nevada, in 2001 that would not only attract tourists but also serve as an actual training area for a real expedition to the Moon. The base is designed to feature a closed-loop habitat in which as many as ten people will live for six months to a year while being studied both physiologically and psychologically.

Space, as Caltech scientist and former Jet Propulsion Laboratory director Bruce C. Murray puts it, is a reflection of Earth. In that regard, political acumen as much as technological prowess defines the space age. As we have seen, NASA was laying the groundwork for a manned mission to the Moon well before President Kennedy decided that doing so would score an unprecedented political victory against the Soviet Union, but his challenge gave the technology the political boost it needed to make it a reality. The result, of course, was the most daring and audacious feat of human exploration in history. Apollo and the manned and unmanned missions of the space program proved that science and technology can readily answer history's call.

Indeed, engineers at what was to become NASA's Lewis Research Center in Cleveland, Ohio, began studying interplanetary propulsion systems as early as 1957 with the stated intention of sending an expedition to Mars. A low-profile "Mars underground" at several of the space agency's centers, as well as at contractors such as Boeing and Lockheed, came up with highly advanced spacecraft for expeditions to the Red Planet. One design was for a five-hundred-foot-long cruiser that could have been lifted out of Arthur C. Clarke's fertile imagination. Two of the behemoths, each carrying a lander, were said to be capable of getting people to Mars by 1982 at a cost of $29 billion. But with the 1971 decision to develop the shuttle, the manned Mars mission officially evaporated.

As the twentieth century turned into the twenty-first, and as the beginning of the move to space took hold in reality, as well as in the collective imagination, compelling reasons for making the old dream come true began to become clear. One reason, a subtle one, is that the inevitable migration off Earth can be accomplished only by close cooperation among many countries and eventually, some futurists believe, by a true planetary civilization. A shared mission of that magnitude would inhibit physical conflict and, by definition, promote cooperation.

Another reason to send explorers and then immigrants to space is the human need for adventure. With most of Earth pretty well charted and occupied, abandoning exploration elsewhere would be soul-deadening, in the view of many of our best thinkers, including Isaac Asimov, Margaret Mead, and Freeman Dyson. It is manifest destiny, they have said, because there is an imperative to nurture humanity's spirit while spreading its seed.

That brings us to the most compelling reason of all: survival of the species. Earth is itself a very "seaworthy" spaceship. Yet the potential for catastrophe is the constant companion of any vessel at sea. It is therefore prudent to have a lifeboat. An asteroid the size of the one that is thought to have done in the dinosaurs, or larger, could do the same to humankind. Astronomers take the threat seriously enough to have formed NEO—for Near-Earth Objects—to study the situation in the closest possible detail. Runaway global warming, an uncontrollable pandemic, or a large-scale nuclear war are only three other possibilities among several.

Far from being superfluous and wasteful, then, the infinite journey to other worlds is the only rational course. As the story on these pages has shown, the odyssey had its origin in the ancient dream and was finally begun by Sputnik and all of its descendants, that have taken the people of Earth into a new dimension that once could only be imagined: from the Sea of Tranquillity to the windswept plains of Mars to Europa's ice-encrusted ocean to the edge of the solar system and beyond.

As you read this, Voyager 1, which describes you and bears greetings from you and others of the race that made and launched it, is heading toward the constellation Hercules. Its sister, carrying the same message, is streaking to a place in the neighborhood of Sagittarius. Pioneers 10 and 11 are taking similar messages to other places very far away. The journey has finally begun, and there is no turning back.

—William E. Burrows

Ray Bradbury, author

Why does NASA exist?

Why do *we* exist?

Why does life exist upon this strange and lonely planet?

How did we arrive and for what reason?

An age-old question. One that each of us at one time or another has asked. Each time the universe responds with silence.

NASA stands before that silence and probes that mystery.

We stand with NASA in response to the incredible miracle of impossible life on an insensate world.

We move back to a Moon that we wish we had never deserted. We move onward to Mars to establish a base and then a community and finally a miniature civilization on its enigmatic soil.

All this will be done not as a technological feat, a military exercise, or a display of human vanity.

We do it because NASA has realized that the universe that extends for billions of light years in all directions is meaningless unless—

Unless what?

Unless there are observers and caretakers of that stunning interstellar display.

The universe *demands* to be noticed, to be seen and dutifully noted. What use all those incredible firework dimensions if no eye fixes and reflects, no brain takes note, no heart moves with passion at the display?

NASA answers the silent cry of the cosmos for recognition.

NASA is the witness and we fellow witnesses to the endless deeps.

NASA's activities are *our* activities. The purpose of life on Earth is to see, to know, and to tell what the cosmos has to offer. Without us human beings, without NASA, the universe would be unseen, unknown, untouched. A mindless abyss of stars asking to be discovered.

So NASA in the coming years will be chief witness and we, as fellow observers, celebrants to the cause. NASA and we have been given the job by genetic accident and surprise of crying Lo! to an unsolvable territory.

NASA and we do this to see, to know, and to prevail. Life one day on Earth will vanish through intemperate heat or cold. We prepare ourselves for the Time of Going.

Away to other worlds, to other stars.

Why? Because we love and value this life and living that has been given to us. Ours is the obligation to see and know and try to understand. First then, the Moon, then threshold Mars, and one far future day to landfall on some world adrift near Alpha Centauri.

Can NASA do this? Can we run tandem with NASA and live forever or a million years, whichever comes first?

We can, we will, we *must!*

Sir Arthur C. Clarke, author

In 1970 it was my privilege to write the epilogue to First on the Moon: A Voyage *with Neil Armstrong, Michael Collins, Edwin E. Aldrin, Jr. Reading it again after thirty years, I find only a few points I would wish to change. So I would therefore like to quote some of the key passages—and then add a 2000 epilogue to my 1970 one.*

Some years ago, a *New Yorker* cartoonist made a profound and witty observation on the nature of life. He drew a primeval beach, lined with the giant ferns that have long since vanished from the earth. Crawling up out of the ocean was a clumsy, archaic fish—a coelacanth, or a close relative—while a few yards to the rear a companion lingered nervously in the deeper water. The intrepid adventurer, already half on dry land, was looking back at his anxious colleague. And he was saying, with a rather patronizing expression: "Because this is where the action is going to be, baby."

From the point of view of any fish, the land is a most unpleasant, hostile place....Yet despite this, life came out of the sea—its ancient birthplace—and conquered the alien land. In so doing, it opened up whole new possibilities of existence which we now take for granted, but which would not have been at all obvious to a fish of human intelligence—could one imagine such a thing—back in the Cambrian period.

But not even the most brilliant and farsighted of fish could have imagined the ultimate consequences of the exploration—and colonization—of the land. He could not have anticipated the rise of new life-forms, with much superior senses and greatly improved ability to manipulate the environment. Long-range vision, and the dexterity of human fingers, could never have evolved in the sea. Nor, in fact, could higher intelligence itself—simply because the benevolent sea does not provide the same challenges as the fierce and inhospitable continents. (Today's intelligent marine animals, the whales and dolphins, are of course all dropouts from the land.)

But, above all, our Paleozoic Leonardo could never have imagined the new technologies which would be discovered and exploited once life had escaped from the all-embracing sea. In particular, the very existence, and the infinite uses, of fire would have been utterly beyond his comprehension. The taming and control of fire is the essential breakthrough which leads to the working

Pages 226–227

This spaceborne radar image (SIR-C/X-SAR) shows part of the vast Namib Sand Sea on the west coast of southern Africa, just northeast of Luderitz, Namibia. In this image, fields of sand dunes are in magenta, the South Atlantic Ocean is in orange along the bottom, and the bright green features in the upper right are rocky hills pointing through the sand sea.

of metals, to prime movers, to electricity—to everything, in fact, upon which civilization depends.

There is no need to pursue the analogy further; the lesson is obvious. When we escape from the ocean of air, we will be moving out into a whole universe of new sensations, experiences, technologies—only a few of which we can foresee today. Zero gravity research, industry, and medicine open up such immense vistas that our descendants will find it impossible to believe that we ever managed without them. Yet the greatest boons from space, it is fairly certain, will come from discoveries still undreamed-of today. Waiting for us beyond the atmosphere are the equivalents, and perhaps the successors, of fire itself.

So much for the opening section of my 1970 epilogue, which then went on to discuss some of the things I expected to see in the three decades that have now ended. Virtually all of these have happened in the fields of space applications and technology, but some (for example: the Orbiter Hilton!) still lie in the future, though perhaps they are already much closer to us in time than the first lunar landing.

And in the much longer run? Here is the conclusion of my original epilogue. I see no reason to change it at the dawn of the new millennium.

Whether we shall be setting forth into a universe which is still unbearably empty, or one which is already full of life, is a riddle which the coming centuries will unfold. Those who described the first landing on the Moon as man's greatest adventure are right; but how great that adventure will really be we may not know for a thousand years.

It is not merely an adventure of the body, but of the mind and spirit, and no one can say where it will end. We may discover that our place in the universe is humble indeed; we should not shrink from the knowledge, if it turns out that we are far nearer the apes than the angels.

Even if this is true, a future of infinite promise lies ahead. We may yet have a splendid and inspiring role to play, on a stage wider and more marvelous than ever dreamed of by any poet or dramatist of the past. For it may be that the old astrologers had the truth exactly reversed when they believed that the stars controlled the destinies of men.

The time may come when men control the destinies of stars.

Carl Sagan, astronomer

Ann Druyan, writer-producer and widow of Sagan: *The last professional thing that Carl did, only days before he died in 1996, was to dictate a letter. He had been invited by Vice President Gore to attend a meeting at the White House on the future of space, of NASA, and of space exploration. The doctors were unequivocal that there was no chance that he could go to the meeting. He was so ill that he was depending on oxygen. He took off his oxygen mask and told me that he had to write a letter to the meeting. I took down his words about the future of space exploration. Vice President Gore began the meeting by reading Carl's letter. And within just two days, he was gone.*

I am firmly convinced that this is an opportunity for bipartisan support not seen in thirty years that can positively affect the far future and convey inspirational and educational values for generations to come. It should not be permitted to be sidetracked by extraneous factors or by parochial interests. The concatenations of two putative discoveries of life on Mars, the apparent discoveries of subsurface oceans on Europa, the finding of dozens of Earthlike new planets of other stars, spectacular new findings of subkilometer life on Earth, the launch of NEAR [Near Earth Asteroid Rendezvous], and the forthcoming launch of Cassini represent a military-free exploration of the solar system that is unprecedented and that appeals to the fascination of the search for extraterrestrial life that so transfixes our children. There should be no backing off from these missions and their objectives in the next ten years.

—Carl Sagan

Homer H. Hickam Jr., author

Most of the interesting work in space over the next few decades will be accomplished by robotic spacecraft, although various countries will continue human spaceflight, primarily for research. During this period, robots will explore the planets and moons of the solar system and primitive forms of life will be discovered. Toward the middle part of the twenty-first century, there will be dramatic improvements in propulsion technology using fission, fusion, or antimatter drives. For the first time, many launch companies will be able to reach space with large payloads. The emphasis in space exploration will then switch from robots to humans. The first ordinary people will begin to travel into space, principally for the purposes of tourism or work, such as the operation of mining machinery on the Moon. It won't take long after that before all the people of Earth will start to see the solar system as a place to live and work.

The true story of spaceflight will not begin, however, until it is precipitated by governmental tyranny, ecological disaster, or religious impulse. A brave people will need to escape, and it won't much matter where they go as long as they can get away. Space will beckon, a place where they can be free and start anew without the restrictions of their old planet. History will then record the most audacious period of adventure, exploration, and migration ever known. It will be both a glorious and a tragic time, when heroic men, women, and children conquer the most forbidding frontier ever known. Where people go, government and civilization will surely follow. In two centuries, colonies will exist across the solar system and humankind will be taking a fix on its first journey to the nearest stars. It will be later recorded that while the seed of Earth prepared to spread across the cosmos, the universe held its breath.

Daniel S. Goldin, NASA administrator

The twenty-first century holds tremendous promise for humankind's quest to explore the depths of space. Within the next twenty-five to fifty years, space travel will be an inexpensive and routine part of our lives, and the safety of low Earth orbit flights will match that of today's long-haul jets.

This will allow hundreds of thousands of people a year to travel into orbit. That experience will inspire us to develop the technologies to send humans to Mars safely within next fifty years. We will also send robotic explorers to the outer planets, including aquabots that will plunge deep into Europa's ocean.

Those deep-space explorers will be biologically inspired machines that land on passing asteroids and comets. They will mine the raw materials they need from their hosts and convert them into advanced structures with new capabilities before heading deeper into space.

To accurately track and effectively communicate with the spacecraft we send across the universe, we will develop an Interplanetary Internet. This network will also provide a complete virtual presence in our solar system to any computer user anywhere in the world.

Our investment today in the technology triangle—biotechnology, nanotechnology, and information technology—will result tomorrow in self-sensing, self-replicating, self-repairing systems that will think, learn, and adapt to dynamic environments.

These amazing systems will be partners with our human astronauts, taking over basic activities and allowing human explorers to do what they do best— think, create, and evaluate.

The technological advances we make will also help us solve many of the health challenges we face in space and on Earth. This, in turn, will allow astronauts to explore the depths of the universe and solve the age old question: Is life unique to Earth?

An Apollo 11 astronaut's boot print in the lunar soil serves as a delicate, yet long-lasting, symbol of humankind's ever-present urge to explore the unknown.

INDEX

ACKNOWLEDGMENTS

Discovery Channel Publishing wishes to thank a number of individuals whose support and contributions made *The Infinite Journey: Eyewitness Accounts of NASA and the Age of Space* possible:

Many special thanks go to the NASA Millennium Committee for the opportunity to collaborate on the conception and creation of this unique millennium project.

Ray Castillo	Ted Nakata
Paula Cleggett	Leslie Tagg
Bobbie Faye Ferguson	Evelyn Thames
Lori Garver	Bertram Ulrich
Alan Ladwig	Lorraine Walton
Roger D. Launius	Brian Welch

Thanks also go to the following people at the National Aeronautics and Space Administration: *NASA Headquarters, Washington, D.C.:* Daniel S. Goldin, NASA Administrator; Peggy Wilhide, Associate Administrator for Public Affairs; Deborah J. Rivera, Lead, New Media, Office of Public Affairs; Jane Odom and Mark Kahn, Archivists, NASA Headquarters History Office; William Bierbower, Office of General Counsel; Carlos Fontanot, Public Affairs Officer; *Lyndon B. Johnson Space Center, Houston, Texas:* William A. Larsen, Assistant to the Director, Information Systems Directorate; Rebecca Wright, Johnson Space Center Oral History Program Manager, Signal Corporation; Mike Gentry, Public Affairs Liaison for the Media Resource Center; Gary Seloff, Manager of Information Systems; Mary N. Wilkerson, Film Repository Supervisor; Jody Russell, Still Photo/Video Librarian; Glen E. Swanson, Historian; Sherri T. Jurls, Newsroom Coordinator, Public Affairs Office; *NASA Goddard Space Flight Center, Greenbelt, Maryland:* Visualization Analysis Lab; Dave Williams, Planetary Scientist, National Space Science Data Center.

Finally, thanks go to Phil Unetic of 3r1 Group; Jane A. Martin; Kathy Rosenbloom of Random House; Pat Goley, Cynthia Richards, and their colleagues at Professional Graphics; Tony Reichhardt; Ronald Beck and Raymond A. Byrnes of the U.S. Geological Survey; Michael Hahn; Ed Harrison; Lola Morrow; Amanda Stowe; and Heather Quinlan.

TEXT & ILLUSTRATION CREDITS

Text Credits

All contributions appearing in *The Infinite Journey: Eyewitness Accounts of NASA and the Age of Space* have been taken from interviews conducted by Discovery Communications, Inc. or authored by participants expressly for the book, with the exception of the following sourced entries:

Page 45, Maxime A. Faget: from a Johnson Space Center Oral History Project interview conducted by Jim Slade, in Houston, Texas, June 18 & 19, 1997; **p. 46,** Alan B. Shepard Jr.: from a Johnson Space Center Oral History Project interview conducted by Roy Neal in Pebble Beach, Florida, February 20, 1998; **p. 48,** M. Scott Carpenter: from the film *In Search of Liberty Bell 7* (The Discovery Channel, 1999); **pp. 49 & 50,** Virgil "Gus" Grissom: from the "Pilot's Flight Report" by Virgil I. Grissom, part of a document on Mercury-Redstone 4 prepared by the Manned Spacecraft Center (http://www.hq.nasa.gov/office/pao/History/MR4/contents. htm); **p. 51,** Lowell Grissom: from the film *In Search of Liberty Bell 7* (The Discovery Channel, 1999); **p. 53,** M. Scott Carpenter: from a Johnson Space Center Oral History Project interview conducted by Roy Neal in Vail, Colorado, January 27, 1999; **p. 54,** John H. Glenn Jr: from the "Brief Summary of MA-6 Orbital Flight" by John H. Glenn Jr., 1962; **p. 55,** John H. Glenn Jr.: from a Johnson Space Center Oral History Project interview conducted by Sherree Scarborough in Houston, Texas, August 25, 1997; L. Gordon Cooper Jr.: from a Johnson Space Center Oral History Project interview conducted by Roy Neal in Pasadena, California, May 21, 1998; Sam T. Beddingfield: excerpt copyright © *St. Petersburg Times,* 1990; **pp. 62 & 63,** James A. McDivitt: from a Johnson Space Center Oral History Project interview conducted by Doug Ward in Elk Lake, Michigan, June 29, 1999; **p. 64,** mission commentary: Gemini-Titan 4 Mission Transcript, NASA History Office, NASA Headquarters, Washington, D.C.; **p. 65,** Edward H. White: *Newsweek,* June 21, 1965, p. 24; **p. 66,** David R. Scott: Postflight quote made on March 19, 1966; **p. 69,** Robert C. Seamans Jr.: Dr. Robert Seamans, Oral History Interview, May 13, 1988, Glennan-Webb-Seamans Project for Research in Space History, National Air and Space Museum; mission commentary transcript courtesy of Dr. David G. Fisher, Associate Professor of Physics and Astronomy, Lycoming College, Pennsylvania; **p. 74,** Roger B. Chaffee: from C. Donald Chrysler, *On Course to the Stars: The Roger B. Chaffee Story* (Grand Rapids: Kregal Publications, 1968); **p. 76,** Virgil I. "Gus" Grissom: 1966, NASA History Office, NASA Headquarters, Washington, D.C.; Eugene F. Kranz: from a Johnson Space Center Oral History Project interview conducted by Rebecca Wright in Houston, Texas, January 8, 1999; **p. 77,** George E. Mueller: from a Johnson Space Center Oral History Project interview conducted by Summer Chick Bergen in Kirkland, Washington, August 27, 1998; **p. 80,** William A. Anders: from a Johnson Space Center Oral History Project interview conducted by Paul Rollins in Houston, Texas, October 8, 1997; **p. 81,** Michael Collins: from a Lyndon B. Johnson Space Center Oral History Project interview conducted by Michelle Kelly, Houston, Texas, October 8, 1997; **p. 83,** Archibald MacLeish: from *The New York Times,* December, 1968; **p. 84,** President John F. Kennedy: from an address to Congress delivered May 25, 1961; **p. 86,** Michael Collins: from Edgar M. Cortright, ed., *Apollo Expeditions to the Moon,* NASA SP-350 (Washington, DC: U.S. Government Printing Office, 1975); Geneva B. Barnes: from Glen E. Swanson, ed., *Before This Decade Is Out: Personal Reflections on the Apollo Program* (Washington, DC: U.S. Government Printing Office, 1999); **p. 88,** Neil A. Armstrong: from http://news.uns. purdue.edu/UNS/html4ever/9903.Armstrong.apollo.html; **p. 89,** Buzz Aldrin: *top right,* from the Apollo 11 post-flight press conference, August 12, 1969; *top left,* from Edgar M. Cortright, ed., *Apollo Expeditions to the Moon,* NASA SP-350 (Washington, DC: U.S. Government Printing Office, 1975); Neil A. Armstrong: from the Apollo 11 technical crew debriefing, July 31, 1969; **pp. 89 & 90,** Michael Collins: from the Apollo 11 post-flight press conference, August 12, 1969; **pp. 94, 95, 96 & 97,** Eugene F. Kranz: from a Lyndon B. Johnson Space Center Oral History Project interview conducted by Roy Neal in Houston, Texas, April 28, 1999; **pp. 100 & 101,** Eugene A. Cernan: from the *Apollo Lunar Surface Journal,* copyright © 2000 Eric M. Jones (http://www.hq.nasa.gov/office/pao/History/alsj/); **p. 102,** Harrison H. Schmitt: from an interview appearing in "Silver Moon," an article on the twenty-fifth anniversary of Apollo 17 by Kelly Humphries of Johnson Space Center, 1997 (http://www.jsc.nasa.gov/pao/apollo17/); **p. 103,** Harrison H. Schmitt: from Glen E. Swanson, ed., *Before This Decade Is Out: Personal Reflections on the Apollo Program* (Washington, DC: U.S. Government Printing Office, 1999); **pp. 102 & 103,** Eugene A. Cernan: courtesy of *The Orange County Register,* July 2, 1999; **p. 110,** Charles "Pete" Conrad:

NASA press conference, February 28, 1973; **pp. 115 & 117,** Thomas P. Stafford: from a Johnson Space Center Oral History Project interview conducted by William Vantine in Houston, Texas, October 15, 1997; **p. 117,** Aleksei A. Leonov: quote copyright © Reuters Ltd. 1999; **p. 125,** Frank Jordan: excerpt courtesy of Tony Reichhardt, from an April 1999 interview; **p. 128,** Hans-Peter Biemann: excerpt revised slightly (with author's permission) from the original version appearing in *The Vikings of '76,* copyright © 1977 by Hans-Peter Biemann; **p. 139,** Henry Casselli: from an interview conducted by Bertram Ulrich, Curator, NASA Art Program; permission for use granted by Artrain USA; **p. 140,** President Ronald Reagan: from an address to NASA employees and families of the *Challenger* crew, delivered January 31, 1986; **p. 142,** Richard L. McAfee: portion of a letter originally appearing on thinkquest.org, reprinted with permission by Richard L. McAfee; **p. 162,** Story Musgrave: from an August 1994 interview by Nina L. Diamond, appearing in *Omni* magazine; **pp. 194 & 196,** Vladimir Georgievich Titov: from a Phase 1 International Space Station Oral History Project (Johnson Space Center) interview conducted in Houston, Texas, on July 21, 1998. Interviewers: Rebecca Wright, Paul Rollins, and Franklin Tarazona; **p. 200,** Norman E. Thagard: from a Phase 1 International Space Station Oral History Project (Johnson Space Center) interview conducted in Houston, Texas, on September 16, 1998. Interviewers: Rebecca Wright, Paul Rollins, and Carol Butler; **p. 219,** John Pearl: from an Associated Press article by Paul Recer, October 13, 1998. **p. 225,** Ray Bradbury: Copyright © 2000 Ray Bradbury for the NASA Art Program; **p. 228,** Sir Arthur C. Clarke: Epilogue from *First on the Moon: A Voyage with Neil Armstrong, Michael Collins, Edwin E. Aldrin Jr.* (New York, NY: Little Brown & Co.). Reprinted by permission of the author and the author's agents, Scovil Chichak Galen Literary Agency, Inc.; **p. 229,** Carl Sagan: excerpt of letter to Vice President Gore courtesy of Ann Druyan.

Illustration Credits

Illustrations appearing in *The Infinite Journey: Eyewitness Accounts of NASA and the Age of Space* are sourced by origin and catalog number.

Images listed with the prefix "Courtesy of NASA" may be found on http://photojournal.jpl.nasa.gov.

Courtesy of NASA/JPL/Caltech
Page 6: *top,* PIA01765, *bottom,* PIA02751; **pp. 8–9:** PIA00014; **p. 34:** PIA01143; **p. 135:** PIA01321; **pp. 148–149:** PIA00104; **p. 150:** PIA00378; **p. 157:** PIA01486; **p. 159:** *bottom left,* PIA01489; *top right,* PIA01969; **p. 161:** PIA01142; **p. 174:** PIA00478; **p. 177:** PIA01191; **pp. 178–179:** PIA00716; **p. 188:** PIA01528; **p. 189:** PIA02403; **pp. 206–207:** PIA01238; **pp. 212–213:** PIA01466; **p. 212:** PIA00908; **p. 213:** PIA01574; **p. 214:** PIA01551; **pp. 214–215:** PIA01424; **pp. 226–227:** PIA01856.

Courtesy of NASA/JPL/Caltech/DLR
Pages 178–179: PIA00502.

Courtesy of NASA/JPL/Caltech/Goddard Space Flight Center
Page 209: PIA01337; **p. 221:** PIA02053.

Courtesy of NASA/JPL/Caltech/Malin Space Science Systems
Pages 218–219: PIA02066.

Courtesy of NASA/JPL/Caltech/United States Geological Survey
Page 1: PIA00342; **pp. 118–119:** PIA00161; **p. 127:** PIA00420; **p. 128:** PIA00153; **p. 129:** PIA00571; **p. 131:** PIA00006; **p. 151:** PIA00307; **p. 158:** PIA00010; **p. 175:** *top left,* PIA00007.

Courtesy of NASA/JPL/Caltech/PIRL,University of Arizona
pp. 178–179: *left to right,* PIA02309, PIA01298; **p.179:** *bottom right,* PIA02505; **Back cover:** PIA01667.

Discovery Virtual Library, Discovery Communications, Inc.
Page 51: Liberty Bell 7.040 copyright © 1999 Discovery.

NASA Ames Research Center
Images may be found on http://ails.arc.nasa.gov.
Page 6–7: AC73-4128; **p. 20:** G69-0901; **pp. 70–71:** G69-44-6552; **p. 123:** AC85-0760-4; **p. 130:** A76-1011-1-65; **p. 153:** AC71-8744; **p. 154:** AC72-1338; **p. 155:** AC73-9319; **p. 160:** A81-7028; **p. 190:** AC96-0110-1.

NASA Goddard Space Flight Center

Images may be found on http://www.gsfc.nasa.gov.

Page 124: MTVS4187-45; **p. 125:** *top left*, MTVS4265-52, *bottom right*, P-4732; **p. 169:** DIRBE; **p. 170:** FIRAS; **p. 171:** DMR; **pp. 180–181:** http://rsd.gsfc.nasa.gov/rsd/images/-HugoPersp.html; **p. 183:** GOES full disk 0001010845.jpg; **p. 217:** MRPS88217.

NASA Johnson Space Center

Images may be found on http://nix.nasa.gov.

Cover: STS064-217-008, **Page 2:** *top*, S69-39799, *bottom*, AS11-44-6609; **p. 3:** G69-44-6552; **p. 11:** STS088-355-015; **pp. 14–15:** S95-13884; **p. 28:** S61-01250; **p. 29:** S74-33007; **p. 32:** 69-HC-894; **p. 33:** AS17-134-20384; **p. 37:** AS08-13-2344; **p. 42:** S88-31375; **p. 45:** S88-31374; **p. 47:** S61-02711; **p. 49:** S61-02882; **p. 50:** S61-02824; **p. 52:** S62-00303; **p. 53:** S64-14854; **p. 55:** S62-06012; **p. 56–57:** S65-63220; **p. 58:** S65-27492; **p. 61:** S65-29642; **p. 62:** S65-19282; **p. 64:** S65-33533; **p. 65:** S65-30427; **p. 67:** S66-24478; **p. 69:** S66-24403; **p. 72:** S68-21355; **p. 73:** AS11-40-5880; **p. 75:** S67-19770; **p. 76:** S67-21294; **p. 79:** AS08-13-2344; **p. 81:** S68-53015; **p. 82:** AS08-13-2225; **p. 83:** AS08-14-2383; **p. 85:** AS11-40-5903; **p. 87:** *bottom left*, S69-40022, *top right*, AS11-37-5528; **p. 88:** AS11-40-5948; **p. 89:** AS11-40-5868; **p. 90:** *top*, S69-39193, *bottom*, AS11-44-6642; **p. 91:** S70-17433; **p. 93:** S70-34986; **p. 94:** AS13-58-8458; **p. 95:** *top*, AS13-60-8675, *bottom*, AS13-62-8929; **p. 96:** S70-35638; **p. 97:** S70-35606; **p. 99:** S72-54813; **p. 100:** AS17-140-21496; **p. 101:** AS17-147-22526; **p. 102:** AS17-148-20380; **p. 103:** *bottom left*, AS17-148-22727, *top right*, S72-55420; **pp. 104–105:** SL4-150-5074; **p. 109:** SL3-114-1682; **p. 111:** SL4-150-5062; **p. 113:** *bottom left*, SL3-115-1837, *top right*, S74-23458; **p. 115:** AST-05-298; **p. 116:** S75-22770; **p. 117:** S74-20831; **p. 121:** S66-17594; **p. 137:** S81-30503; **p. 141:** ST51(L)002; **p. 142:** S86-38989; **p. 143:** STS51L(S)029; **p. 145:** STS026-46-048; **p. 146:** S88-42100; **p. 147:** S88-25804; **p. 163:** STS061-95-028; **p. 173:** S91-50686; **p. 175:** *bottom right*, S90-53333; **p. 193:** STS088-705-070; **p. 195:** STS063(5)005; **p. 196:** STS063-715-029; **p. 197:** *top left*, STS063-711-080, *bottom right*, STS063-710-005; **p. 200:** S94-47050; **p. 202:** S99-00883; **p. 203:** STS088-702-024; **p. 204:** STS088-703-019; **p. 205:** STS088-341-015; **p. 231:** AS11-40-5878.

NASA Kennedy Space Center

Images may be found on http://www.ksc.nasa.gov.

Pages 4-5: KSC-95EC-0395; **p. 43:** 63-MA9-137; **p. 86:** KSC-69PC-0295; **p. 117:** *top right*, KSC-75P-0134;

pp. 132–133: KSC-81PC-0136; **p. 138:** S80-40378; **p. 146:** *top left*, KSC-88PC-1176; **p. 199:** KSC-95EC-0948; **p. 211:** KSC-96PC-1147.

NASA Langley Research Center

Images may be found on http://lisar.larc.nasa.gov.

Page 17: EL-1996-00190; **p. 40:** EL-1996-00089.

NASA Marshall Space Flight Center

Images may be found on http://mix.msfc.nasa.gov.

Page 25: 9131496; **p. 35:** 9400251; **p. 80:** 6872176; **p. 107:** 7025449; **p. 110:** 7040570; **p. 112:** 7200-405; **p. 164:** 9263351; **p. 165:** *top right*, 8218871; **p. 201:** 9701025.

NASA, A. Fruchter and the ERO Team

Images may be found on http://oposite.stsci.edu/pubinfo.

Page 165: STScI-PRC00-07; **p. 167:** STScI-PRC00-08.

NASA Headquarters

Images may be found on http://www.nasa.gov.

p. 21: 78-H-136; **p. 22:** 73-H-787; **p. 31:** S69-51308; **p. 46:** S61-02385; **p. 54:** 62-MA6-163; **p. 59:** 69-HC-190; **p. 63:** S65-33532; **p. 77:** 67-H-141; **p. 139:** 82-H-301 courtesy of NASA Art Program; **p. 185:** *top left*, G74-7446, *bottom right*, SA69-15356.

NASA, The Hubble Heritage Team

Images may be found on http://oposite.stsci.edu/pubinfo.

p. 38: STScI-PRC00-12; **p. 166:** STScI-PRC00-06; **p. 223:** STScI-PRC00-10.

The White House, Washington, D.C.

p.12: Official White House photograph: Ralph Alswang, The White House.

NASA Stennis Space Center

Images may be found on http://www.ssc.nasa.gov/about/history.

pp. 18–19: 67-2189; **p. 27:** 67-620.

United States Geological Survey/EROS Data Center

p. 187: *left*, 1173028007314990, *right*, L71161028_02819991116.

Discovery Communications, Inc.

John S. Hendricks
Founder, Chairman, and Chief Executive Officer

Judith A. McHale
President and Chief Operating Officer

Judy L. Harris
*Senior Vice President and General Manager
Consumer & Educational Products*

Discovery Channel Publishing

Natalie Chapman
Vice President, Publishing

Rita Thievon Mullin
Editorial Director

Michael Hentges
Design Director

Mary Kalamaras
Senior Editor

Maria Mihalik Higgins
Editor

Rick Ludwick
Managing Editor

Christine Alvarez
Business Development

Jill Gordon
Marketing Manager

Staff for The Infinite Journey

Mary Kalamaras, *editor*
Michael Hentges, *design director*
Jane A. Martin, *picture editor*
Tony Reichhardt, *interviewer/consultant*
Irene K. Brown, Linda D. Voss, Margaret A. Weitekamp,
Karen Schwarz, and Robert N. Wold, *interviewers*
Amanda Stowe, *researcher*
Julia Duncan, *copy editor*
Mary Mayberry, *researcher*
Barbara Klein, *indexer*

Book and cover design
Phillip Unetic, 3r1 Group, Willow Grove, PA

Discovery Communications, Inc., produces high-quality nonfiction television programming, interactive media, books, films, and consumer products. Discovery Networks, a division of Discovery Communications, Inc., operates and manages the Discovery Channel, TLC, Animal Planet, Travel Channel, and the Discovery Health Channel.

Discovery Books and the Discovery Books colophon are trademarks of Discovery Communications, Inc.

Library of Congress Cataloging-in-Publication Data on file with the Library of Congress

ISBN 1-56331-924-1

Discovery Communications website address:
www.discovery.com
Random House website address:
www.randomhouse.com

Printed in the United States on acid-free paper
10 9 8 7 6 5 4 3 2 1
First Edition